METHODS IN MOLECULAR BIOLOGY

Series Editor
John M. Walker
School of Life and Medical Sciences
University of Hertfordshire
Hatfield, Hertfordshire, UK

For further volumes:
http://www.springer.com/series/7651

For over 35 years, biological scientists have come to rely on the research protocols and methodologies in the critically acclaimed *Methods in Molecular Biology* series. The series was the first to introduce the step-by-step protocols approach that has become the standard in all biomedical protocol publishing. Each protocol is provided in readily-reproducible step-by-step fashion, opening with an introductory overview, a list of the materials and reagents needed to complete the experiment, and followed by a detailed procedure that is supported with a helpful notes section offering tips and tricks of the trade as well as troubleshooting advice. These hallmark features were introduced by series editor Dr. John Walker and constitute the key ingredient in each and every volume of the *Methods in Molecular Biology* series. Tested and trusted, comprehensive and reliable, all protocols from the series are indexed in PubMed.

Spectral and Imaging Cytometry

Methods and Protocols

Second Edition

Edited by

Natasha S. Barteneva

Department of Biology, School of Sciences and Humanities, Nazarbayev University, Astana, Kazakhstan;
Brigham Women's Hospital, Harvard University, Boston, MA, USA

Ivan A. Vorobjev

Department of Biology, School of Sciences and Humanities, Nazarbayev University, Astana, Kazakhstan

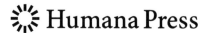

Editors
Natasha S. Barteneva
Department of Biology
School of Sciences and Humanities
Nazarbayev University
Astana, Kazakhstan

Brigham Women's Hospital
Harvard University
Boston, MA, USA

Ivan A. Vorobjev
Department of Biology
School of Sciences and Humanities
Nazarbayev University
Astana, Kazakhstan

ISSN 1064-3745   ISSN 1940-6029   (electronic)
Methods in Molecular Biology
ISBN 978-1-0716-3022-8   ISBN 978-1-0716-3020-4   (eBook)
https://doi.org/10.1007/978-1-0716-3020-4

This Humana imprint is published by the registered company Springer Science+Business Media, LLC, part of Springer Nature.
The registered company address is: 1 New York Plaza, New York, NY 10004, U.S.A.

Dedication

In memory of Aleksandra Bergman-Evstafieva: "The crocodile cannot turn its head. Like all science, it must always go forward with all-devouring jaws" – Pyotr Kapitsa

Preface

Six years ago, we published a volume on *Imaging Flow Cytometry* in the *Methods in Molecular Biology* series [1] and would now like to extend the presentation of the capabilities of modern cytometry development in this series from a broader perspective.

In recent years, flow cytometry has demonstrated significant progress in technology (data acquisition) and data analysis. Improvement of flow cytometry is now directed to overcoming limitations of the traditional flow cytometry technology in the number of colors (labels) that could be detected simultaneously and the inability to capture images of each cell (object) during analysis. Two branches are emerging: imaging flow cytometry (IFC) is gaining popularity, and spectral flow cytometry (SFC) instruments became commercially available just several years ago. This volume aims to present an overview of spectral cytometry, recently developed protocols in the IFC area, and several protocols developed by using FlowCam – a special imaging cytometer whose potential in basic research is still underestimated.

Multi-laser flow cytometers made it possible to detect up to 15–18 colors [2–4]; however, the compensation tables, in this case, become critically large, and detection is usually limited only to relatively bright populations. It means that despite the many colors used, not all subpopulations, for example in bone marrow probes, could be distinguished unambiguously. This limitation has recently been overcome by introducing spectral instruments equipped with PMT arrays (altogether, it has up to 186 PMTs), with each detector assigned to the narrow part of the spectrum. The development of special spectral unmixing algorithms for data analysis improved the discrimination of the populations stained with dyes having similar fluorescent spectra compared to the spectral compensation algorithms in conventional cytometry. Spectral cytometry allowed to extend of the multicolor panel beyond 40 colors [5]. The most prominent advantage of SFC is its ability to analyze autofluorescent spectra in detail. This feature gives new capabilities to the analysis of heterogeneous populations of blood cells [6–8] and will become an indispensable tool for the analysis of algae and cyanobacteria.

Another problem of conventional flow cytometry is the inability to take pictures of individual cells. Information about objects in the flow cell of a conventional cytometer is limited by its light scatter properties. For example, doublets cannot always be resolved from singlets making analysis of the rare events extremely difficult. Visualization of individual cells in the brightfield mode and in fluorescent channels during population analysis was resolved by introducing Imagestream-100 and later Imagestream X and Mark II models (Amnis) with the time-delayed camera(s). Imagestream instrument (Amnis) was developed for the analysis of relatively small cells similar to those in the conventional flow cytometry and used the cuvette 250 μm in diameter and 875 μm in depth. It allows to obtain high-resolution images of cells during flow experiments in multiple fluorescent channels with high sensitivity. The image-enabled intelligent high-speed cell sorting of single cells is under development [9, 10].

A separate line of evolution of the flow instruments resulted in the introduction of FlowCam, designed for the analysis of relatively large objects. FlowCam was initially developed for the analysis of phytoplankton and some zooplankton and later became a useful instrument for the evaluation of aggregated particles in pharma industries. Specific features of the FlowCam are interchangeable objectives with different magnification and flow cells of different diameters allowing analysis of planktonic organisms up to 600 µM. However, FlowCam is mainly addressed to the analysis in the brightfield mode using a color camera, and its fluorescent capabilities are limited by not more than two channels. The image gallery is created by this instrument offline using special software.

The basic knowledge and techniques of IFC have been well documented in our previous volume (2016). Two aspects are considered in the current volume – the development of rapidly emerging spectral cytometry and some applications of imaging flow cytometry.

This new volume is organized into three parts. The first part provides an introduction to state-of-the-art spectral cytometry. In the first chapter, the authors review a relatively short history of spectral cytometry development and discuss its advantages compared to conventional cytometry. The second chapter demonstrates the possibility of discriminating different phytoplankton species based on cell autofluorescence and provides a detailed protocol of virtual filtering. At first glance, the absence of chapters on a detailed description of the new SFC techniques may seem like an oversight for a volume having "Spectral Flow Cytometry" in the title. However, given the rapidly evolving nature of SFC and the recent introduction of new instruments, we expect special volumes dedicated to this novel technology will likely be forthcoming in the next 2–3 years. The second part describes several novel applications of imaging flow cytometry using Imagestream instrumentation for semi-quantitative and quantitative analysis in different experimental models. Chapters reflect ongoing IFC advances in quantitative analysis of pathogens (*Legionella pneumophila*), analysis of multinuclearity, quantitative biodosimetry, and autophagy protocols. Moreover, methods for quantification of specific organelles, such as Golgi complex and inflammasomes have been added to the current volume. The third part contains detailed protocols for handling and using the FlowCam imaging flow cytometer from the supplier and research protocol for the studies of phytoplankton communities.

We are extremely grateful to all authors who provided chapters for this volume during the difficult time of the COVID-19 epidemics. Last but not least, we would like to thank Professor John Walker, Editor of the *Methods in Molecular Biology* series, for his unlimited guidance and help.

We believe that this volume will be a valuable source for a wide audience looking for new approaches in cytometry. The development of methods that will become instrumental in spectral cytometry is continuing, whereas the IFC is becoming a matured cytometry method. This is truly a fascinating time to be involved in cytometry, as spectral and imaging cytometry continues to evolve at an amazing speed.

Astana, Kazakhstan   *Natasha S. Barteneva*
                                                                       Ivan A. Vorobjev

References

1. Imaging Flow Cytometry: Methods and Protocols (Methods in Molecular Biology vol.1389), 1st Edition, 2016. Eds. Natasha S. Barteneva, Ivan A, Vorobjev. pp. 308.

2. Moncunill G, Han H, Dobano C, McElrath MJ, De Rosa SC (2014) Pan-leukocyte immunophenotypic characterization of PBMC subsets in human samples. Cytometry A 85: 995–998. https://doi.org/10.1002/cyto.a.22580.

3. https://www.beckman.kz/resources/reading-material/application-notes/18-color-human-blood-phenotyping-flow-cytometry

4. https://www.agilent.com/cs/library/applications/application-osmotic-fragility-novocyte-5994-1029en-agilent.pdf

5. Sahir F, Mateo JM, Steinhoff M, Siveen KS (2020) Development of a 43 color panel for the characterization of conventional and unconventional T-cell subsets, B cells, NK cells, monocytes, dendritic cells, and innate lymphoid cells using spectral flow cytometry. Cytometry 2020:1–7. https://doi.org/10.1002/cyto.a.24288

6. Peixoto MM, Soares-da-Silva F, Schmutz S, Mailhe M-P, Novault S, Cumano A, Ait-Mansour C (2022) Identification of fetal liver stroma in spectral cytometry using the parameter autofluorescence. Cytometry A 2022. https://doi.org/10.1002/cyto.a.24567.

7. Adusei KM, Ngo TB, Alfonso AL, Lokwani R, DeStefano S, Karkanitsa M, Spathies J, Goldman SM, Dearth CL, Sadtler KN (2022) Development of a high-color flow cytometry panel for immunologic analysis of tissue injury and reconstruction in a rat model. Cells Tissues Organs. https://doi.org/10.1159/000524682

8. Heieis GA, Patente TA, Tak T, Almeida L, Everts B (2022) Spectral flow cytometry reveals metabolic heterogeneity in tissue macrophages. BioRxiv. doi: https://doi.org/10.1101/2022.05.26.493548

9. Schraivogel D, Kuhn TM, Rauscher B, Rodríguez-Martínez M, Paulsen M, Owsley K, Middlebrook A, Tischer C, Ramasz B, Ordoñez-Rueda D, Dees M (2022) High-speed fluorescence image–enabled cell sorting. Science 375: 315–320. doi: https://doi.org/10.1126/science.abj3013

10. Salek M, Li N, Chou HP, Sinai K, Jovic A, Jacobs K, Johnsson C, Lee E, Chang C, Nguyen P, Mei J. (2022) Sorting of viable unlabeled cells based on deep representations links morphology to multiomics. Research Square. Preprint. doi: https://doi.org/10.21203/rs.3.rs-1778207/v1

Contents

Contributors

NATASHA S. BARTENEVA • *Department of Biology, School of Sciences and Humanities, Nazarbayev University, Astana, Kazakhstan; Brigham Women's Hospital, Harvard University, Boston, MA, USA; The EREC, Nazarbayev University, Astana, Kazakhstan*

LINDSAY A. BEATON-GREEN • *Consumer and Clinical Radiation Protection Bureau, Health Canada, Ottawa, ON, Canada*

SULTAN BEKBAYEV • *School of Sciences and Humanities, Nazarbayev University, Astana, Kazakhstan*

KAVITA BISHT • *Mater Research Institute, The University of Queensland, Woolloongabba, QLD, Australia*

VERONIKA DASHKOVA • *School of Sciences and Humanities, Nazarbayev University, Astana, Kazakhstan; School of Digital Sciences and Engineering, Nazarbayev University, Astana, Kazakhstan; PhD Program in Science, Engineering and Technology, Nazarbayev University, Astana, Kazakhstan*

WENDY N. ERBER • *Translational Cancer Pathology Laboratory, School of Biomedical Sciences (M504), The University of Western Australia, Crawley, WA, Australia; PathWest Laboratory Medicine, Nedlands, WA, Australia*

MARIA EROKHINA • *Faculty of Biology, Lomonosov Moscow State University, Moscow, Russian Federation*

KATHY A. FULLER • *Translational Cancer Pathology Laboratory, School of Biomedical Sciences (M504), The University of Western Australia, Crawley, WA, Australia*

JONATHAN A. HARTON • *Department of Immunology and Microbial Disease, Albany Medical College, Albany, NY, USA*

RACHEL E. HEWITT • *Department of Veterinary Medicine, University of Cambridge, Cambridge, UK*

HUBERT HILBI • *Institute of Medical Microbiology, University of Zürich, Zürich, Switzerland*

HENRY Y. L. HUI • *Translational Cancer Pathology Laboratory, School of Biomedical Sciences (M504), The University of Western Australia, Crawley, WA, Australia*

DARIO HÜSLER • *Institute of Medical Microbiology, University of Zürich, Zürich, Switzerland*

AIGUL KUSSANOVA • *School of Sciences and Humanities, Nazarbayev University, Astana, Kazakhstan; Core Facilities, Nazarbayev University, Astana, Kazakhstan*

YOUNGHYUN LEE • *Laboratory of Biological Dosimetry, National Radiation Emergency Medical Center, Korea Institute of Radiological and Medical Sciences, Seoul, Republic of Korea; Department of Biomedical Laboratory Science, College of Medical Sciences, Soonchunhyang University, Asan, Republic of Korea*

JEAN-PIERRE LEVESQUE • *Mater Research Institute, The University of Queensland, Woolloongabba, QLD, Australia; Translational Research Institute, Woolloongabba, QLD, Australia*

DMITRY MALASHENKOV • *Department of Biology, School of Sciences and Humanities, Nazarbayev University, Astana, Kazakhstan*

AYAGOZ MEIRKHANOVA • *School of Sciences and Humanities, Nazarbayev University, Astana, Kazakhstan*

ABHINIT NAGAR • *Program in Innate Immunity, Division of Infectious Diseases and Immunology, School of Medicine, University of Massachusetts, Worcester, MA, USA*

EKATERINA PAVLOVA • *Faculty of Biology, Lomonosov Moscow State University, Moscow, Russian Federation*

SVETLANA PETRICHUK • *National Medical Research Center for Children's Health, Laboratory of Experimental Immunology and Virology, Moscow, Russian Federation*

ZIV PORAT • *Flow Cytometry Unit, Department of Life Sciences Core Facilities, Weizmann Institute of Science, Rehovot, Israel*

JONATHAN J. POWELL • *Department of Veterinary Medicine, University of Cambridge, Cambridge, UK*

TATIANA RADYGINA • *National Medical Research Center for Children's Health, Laboratory of Experimental Immunology and Virology, Moscow, Russian Federation*

KATHRYN H. ROACHE-JOHNSON • *Yokogawa Fluid Imaging Technologies, Scarborough, ME, USA*

MATTHEW RODRIGUES • *Luminex Corporation, Seattle, WA, USA*

KI MOON SEONG • *Laboratory of Biological Dosimetry, National Radiation Emergency Medical Center, Korea Institute of Radiological and Medical Sciences, Seoul, Republic of Korea*

DARIA SHAPOSHNIKOVA • *Faculty of Biology, Lomonosov Moscow State University, Moscow, Russian Federation*

NICOLE R. STEPHENS • *Yokogawa Fluid Imaging Technologies, Scarborough, ME, USA*

JOSHUA TAY • *Mater Research Institute, The University of Queensland, Woolloongabba, QLD, Australia*

ADIL TEMIRGALIYEV • *School of Sciences and Humanities, Nazarbayev University, Astana, Kazakhstan*

MADINA TLEGENOVA • *National Laboratory Astana, Nazarbayev University, Astana, Kazakhstan*

HELEN C. TURNER • *Center for Radiological Research, Columbia University Irving Medical Center, New York, NY, USA*

BRADLEY VIS • *Department of Veterinary Medicine, University of Cambridge, Cambridge, UK*

IVAN A. VOROBJEV • *Department of Biology, School of Sciences and Humanities, Nazarbayev University, Astana, Kazakhstan; National Laboratory Astana, Nazarbayev University, Astana, Kazakhstan; A.N. Belozersky Insitute of Physico-Chemical Biology, Lomonosov Moscow State University, Moscow, Russian Federation; Biological Faculty, Lomonosov Moscow State University, Moscow, Russian Federation*

QI WANG • *Center for Radiological Research, Columbia University Irving Medical Center, New York, NY, USA; Radiation Oncology, Columbia University Irving Medical Center, New York, NY, USA*

AMANDA WELIN • *Division of Inflammation and Infection, Department of Biomedical and Clinical Sciences, Linköping University, Linköping, Sweden*

RUTH C. WILKINS • *Consumer and Clinical Radiation Protection Bureau, Health Canada, Ottawa, ON, Canada*

INGRID G. WINKLER • *Mater Research Institute, The University of Queensland, Woolloongabba, QLD, Australia*

INBAL WORTZEL • *Children's Cancer and Blood Foundation Laboratories, Departments of Pediatrics, and Cell and Developmental Biology, Drukier Institute for Children's Health, Meyer Cancer Center, Weill Cornell Medicine, New York, NY, USA*

Part I

Spectral Flow Cytometry

Chapter 1

Development of Spectral Imaging Cytometry

Ivan A. Vorobjev, Aigul Kussanova, and Natasha S. Barteneva

Abstract

Spectral flow cytometry is a new technology that enables measurements of fluorescent spectra and light scattering properties in diverse cellular populations with high precision. Modern instruments allow simultaneous determination of up to 40+ fluorescent dyes with heavily overlapping emission spectra, discrimination of autofluorescent signals in the stained specimens, and detailed analysis of diverse autofluorescence of different cells—from mammalian to chlorophyll-containing cells like cyanobacteria. In this paper, we review the history, compare modern conventional and spectral flow cytometers, and discuss several applications of spectral flow cytometry.

Key words Spectral cytometry, Flow cytometry, Fluorescence spectra, Aurora cytometer, Sony spectral analyzer, Autofluorescence, Spectral unmixing, Virtual filtering

1 Introduction

Flow cytometry began its development in the middle of the twentieth century and has established itself as one of the major functional methods widely used by scientists and clinicians. As it developed, flow cytometry in the twenty-first century diverges into the following directions: (1) Conventional flow cytometry and fluorescent activated cell sorting (FACS); (2) Imaging flow cytometry; (3) Spectral flow cytometry (spectral FCM).

Conventional cytometry allows studying the size, granularity, and several fluorescent signals of individual cells or particles at the rate of 1000 events per second. Imaging flow cytometry, a hybrid technology, which combines the principles of flow cytometry and microscopy, allows obtaining an image of each cell and thus collects galleries of images along with light scatter and fluorescent signals. However, its throughput is significantly less than conventional flow cytometry [1]. Spectral FCM, which is based on spectroscopy, made it possible to record the full spectrum of every single cell during measurements and now operates at a rate similar to conventional flow cytometry. Both imaging flow cytometry and spectral

Natasha S. Barteneva and Ivan A. Vorobjev (eds.), *Spectral and Imaging Cytometry: Methods and Protocols*, Methods in Molecular Biology, vol. 2635, https://doi.org/10.1007/978-1-0716-3020-4_1, © The Author(s) 2023

FCM allow sophisticated offline analysis of the specimens. Recent technical advances in multicolor cytometry were focused on detecting and analyzing cellular subpopulations with complex immunophenotypes participating in the immune response to diseases and/or vaccine response [2, 3]. Besides, significant progress in the decomposition of complex fluorescent spectra was introduced by Rosetti and co-authors [4], which could improve spectral unmixing and detection of autofluorescence. It will allow better separation of negative, dim, and positive populations using multicolor labeling.

2   Development of Spectral Flow Cytometry

Wade and colleagues made one of the first attempts to extract full emission spectra during flow cytometry analysis in 1979 [5]. They used a grating spectrograph and projected the spectrum of the fluorescent signal onto TV type Vidicon detector (objects: *Anacyctis nidulans*—chlorophyll and phycobilin, 600–750 nm; 3T3 fibroblasts from Balb/c mice, propidium iodide (PI) and fluorescamine staining) [5]. The recorded signal from individual cells has a very low signal-to-noise (S/N) ratio, and reasonable spectra were obtained only by averaging recordings from hundreds of cells (fibroblasts).

The next configuration of the spectral flow cytometer was based on a photomultiplier tube (PMT) as a detector and grating monochromator. This cytometer had a 10 nm bandwidth spectral resolution, and signal detection was performed when cells were running through the cuvette at a rate of hundreds of events per second [6]. The system eliminated problems of the noisy background using PMT with adjustable gain and offset as a detector (analyzed objects—fixed rat thymocytes, stained with Hoechst 33258) [6]. However, the spectrum measured was only between 400 and 600 nm, not including far-red and infrared (IR) wavelengths.

Buican [7] described in 1990 a "real-time FT spectrometer" that was an interferometer-based spectral detector using PMT with minimal time needed for the recording of the spectrum (only 3.2 μs). However, this instrument was never used as a commercially available cytometer. Subsequently, several more spectral systems were created in an attempt to obtain spectra from short measurements using conventional cytometers in 1990–2000. Thus, Gauci and co-authors described configuration with the prism and 512-element intensified photodiode array based on the FACS IV laser flow cytometer. They analyzed spectra obtained from *Dictyostelium discoideum* spores stained with Cy3, fluorescein isothiocyanate, R-phycoerythrin (R-PE), and calibration beads [8]. This system was relatively slow (operating at 62.5 Hz) and not sensitive enough to show individual spectra of the labeled cells [8]. Further-more, Asbury et al. (1996) [9] were the first group able to obtain the fluorescent spectrum using a standard flow cytometer (Cytoma-tion, USA) with a monochromator attached in front of PMT used as a detector of fluorescent signal. However, this was not a real spectral FCM yet. The monochromator was operating sequentially—for each wavelength (spectral point), 100 events were recorded. Then monochromator was shifted to the adjacent position, recording another 100 events and so on. The overall spectrum (400–800 nm) was built up from measurements made on 20,000 particles.

At the beginning of the multicolor analysis, the sensitivity of flow cytometers and confocal microscopes in the far-red and IR parts of the spectrum was limited by the low sensitivity of PMTs at wavelengths beyond 650 nm [10]. The use of avalanche photodi-ode detectors (APD) led to substantially better S/N performance over the PMT in the red and near-IR spectral regions. Changing conventional PMTs to APD and APD arrays [11, 12] made it possible to achieve reasonable S/N for multichannel detectors using short-time exposures even in near IR (wavelengths up to 800–900 nm) [13]. An alternative type of detector was used by Isailovic and colleagues [14]. Their instrument (single-cell fluores-cence spectrometer) was based on ICCD (intensified charge couple device) detector and used a 5–20 ms exposure time, thus coming close to the real spectral FCM. Using this instrument, they demon-strated that measurement of individual spectra with a spectral reso-lution of 6.5 nm from fluorescently labeled *E. coli* expressing GFP and non-fluorescent apo-subunits of R-PE gives more accurate results compared to the measurement of bulk spectra.

Since the beginning of the twenty-first century, various systems have achieved sufficient sensitivity for recording a spectrum of fluorescent signals from a single cell in a reasonably short time. The next step in the development of spectral flow cytometers became possible when computer speed accelerated and paralleled recording of multiple signals with high frequency was achieved on a standard PC. Rapid registration of fluorescent spectra was done using parallel data recording and digital processing. These instru-ments were based on multidetector arrays, where emission light is split and projected onto the grid of PMTs or APDs. A flow cyt-ometer equipped with 32-channel Hamamatsu multi-anode PMT able to collect spectral information in not more than 5 μs was built in Purdue University Cytometry Laboratory and later patented by Purdue University [15, 16]. This instrument allowed a digitization rate of up to 75,000 complete a 32-channel spectra per second at 14 bits dynamic range for uniformly (in time) presented events. The system was based on an EPICS Elite cell sorter (Beckman Coulter, USA) equipped with argon (488 nm) and HeNe

(633 nm) lasers [16]. This system achieved a speed of 3000 random events per second; however, the sensitivity was lower than that of conventional filter-based detectors.

A similar system based on a modified BD FACSCalibur cytometer equipped with argon-ion laser and 100 W mercury lamp was built by Goddard and co-authors [17] using a grating spectrograph and Hamamatsu CCD array with 80% quantum yield. The spectra analyzed by this instrument were in the range of 500–800 nm. This instrument allowed recording spectra with great linearity, making spectral subtraction to remove background signals from labeled specimens such as Rayleigh scattering, Raman light scatter, and even cell autofluorescence feasible. Also, the sensitivity of the instrument was significantly lower (10–30 times) than that of the conventional cytometer [17].

Alternative spectral cytometry systems used a charge-coupled device (CCD) camera as a detector to measure spectra from single cells and beads [17–19]. In 2012 Nolan's group [20] developed spectral FCM instruments and data analysis algorithms suitable for everyday use. Their two systems were based on FACSCanto equipped with 405 and 488 nm lasers and using EM-CCD (electron-multiplying CCD) detector (11.3 nm resolution in the 500–800 nm range) and Coulter Elite cytometer using 785 nm laser for IR emission (at 3.23 nm resolution in 790–930 nm range). Their spectral flow cytometers used a holographic grating and EM-CCD detector for high-speed spectra detection. Customized software was developed for the spectral unmixing and production of spectra-derived parameters for individual cells.

Instrument calibration and data analysis were very complicated at these early stages of spectral FCM development (circa 2012) [21]. Instrument design was not standardized, requiring thorough spectral calibration for each instrument. Also, different instruments used different data formats, making cross-platform spectral analysis tricky. In the first spectral cytometers, spectral unmixing was performed through the least square unmixing algorithm or indirectly through principal component analysis [22]. Overall comparing the spectral data obtained by different instruments was practically impossible. So far, at that time, the advantages of spectral FCM over conventional multichannel flow cytometry were impossible to use in many applications. The next step was done when commercially available spectral cytometers with standardized parameters appeared.

The system patented by Purdue University was licensed by Sony Inc., which is producing the first-generation commercial spectral cytometry system (sometimes named hyperspectral cytometer)—the Sony SP6800 Spectral Analyzer was announced at the end of 2012 and came to the market in 2014. Also, in 2014 Cytek Biosciences (USA) developed and soon released its Aurora spectral flow cytometer. Nowadays, two companies are concerned with the production of commercial models of spectral cytometers: (1) Sony Biotechnology (spectral cell analyzers SA 3800, SP 6800, ID 7000); (2) Cytek Biosciences (Cytek Aurora and Northern Lights instruments). In summary, recent advances in hardware, detectors, and computer analysis algorithms resulted in commercially available spectral FCM instrumentation.

3   Current Spectral Cytometry Instruments

Modern Sony ID7000 instrument supports up to 7 lasers and can use up to 168 detectors (in 7 laser configuration) covering the spectral range from 360 to 920 nm with ~10 nm resolution. *Specialized InGaAs PMTs are used for efficient capturing of the IR signals.*

Aurora Cytek spectral cytometer measures fluorescence in up to 64 fluorescent channels (in the 5-lasers instrument— 16UV + 16 V + 14B + 10YG + 8R) across the APD detector arrays (Fig. 1). Each channel uses a special bandpass filter with about 10–15 nm bandwidth, reflecting all wavelengths outside of its transmission band. The full spectral range is 400–900 nm. In both types of instruments, lasers excite the specimen sequentially.

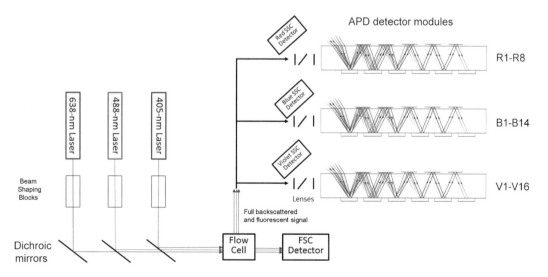

Fig. 1 Three laser Aurora Cytek instrument—optical setup. Fluorescence signal is delivered to the sets of detectors (*V* for violet excitation, *B* for blue excitation, and *R* for red excitation). Notice that SSC signal is measured for each laser, and the number of APD detectors is different. Laser beams are spatially separated at the conventional cytometer. Picture was modified from figures given at Aurora Cytek website (https://cytekbio.com>pages>aurora-cs)

4   Advances and Limitations of Spectral Flow Cytometry

A critical review of the latest advances and remaining problems in spectral FCM was published recently [23]. The essential aspects of spectral FCM are that instrument performance in the case of Cytek Aurora strongly depends on the characteristics of each filter (total—of 64 filters). For example, a thorough check uncovered two out-of-specification filters in the commercial instrument that precluded efficient separation of eFluor450 from BV421 and SB436 [23]. Other issues dealt with laser delay and titration of antibodies. In the case of spectral FCM, titration of antibodies is more complicated because of living and dead cells in the same tube. Authors suggest using live and dead cell markers along with a standard set of CD markers, making titration a multistep process. This process can be described as inversed to FMO (fluorescence minus one) controls used in conventional multicolor cytometry. The sequence of suggested tests for titration is the following: viability dye, major markers like CD45, lineage-specific markers (CD3, CD19, etc.), and finely more specific markers to identify small subpopulations of blood cells [23].

5   Development of Spectral Unmixing Algorithms

The significant advantage of spectral measurements against conventional flow cytometry is its ability to make a detailed comparison of fluorescent spectra from individual cells (objects) in a heterogeneous population. Multiparametric cytometry often has bleed-through problems due to the overlapping spectra of fluorophores. To identify and characterize complex interactions of multiple cell types, it is necessary to analyze a significant number of fluorescent labels simultaneously. Fluorescence signals were initially analyzed as a linear combination of reference spectra with algorithms extracting the weight of individual spectra (linear unmixing) [24]. Identification of heavily overlapping spectra can be performed to a limited extent using the spectral compensation procedure, and instead, spectral unmixing was introduced. Spectral unmixing refers to a group of techniques that attempt to determine how much each fluorophore contributes to the observed emission spectrum. It was initially suggested for microscopy [25] and later applied in flow cytometry [21, 26]. Spectral unmixing in cytometry allows analysis of the simultaneous labeling of cells with several fluorophores and/or fluorescent proteins. Spectral unmixing methods have been developed extensively for the remote sensing analysis of hyperspectral data [27, 28]; however, some key differences make many unmixing algorithms unsuitable for spectral cytometry: (a) the number of fluorophores used for cellular staining is known a priori, though the number of autofluorescent signals can be unknown; (b) remote sensing spectral analysis is focused on blind unmixing of source signals while in spectral cytometry it is possible to use reference spectra to define emission spectral endmembers.

6   Spectral Unmixing Problems

The fluorophores originated from algal photosynthetic apparatus such as PE, APC, and PERCP have broad and overlapping spectra, and to some degree, can be excited by violet laser (405 nm excitation) [29]. Synthetic dyes such as Alexa Fluor and Cyan families are small organic fluorophores that do not exhibit much crossbeam excitation. Most spectral unmixing algorithms cannot separate a signal from background noise or autofluorescence. Autofluorescence is a common, undesired signal arising from endogenous fluorophores contained in the cells or extracellular matrix (i.e., NAD(P)H, flavine adenine nucleotide (FAD), lipids, collagen, elastin, and other common fibrous proteins, porphyrins) [30] often with wide emission spectra [31]. One of the major endogenous fluorophores inside cells is a mitochondrial NADH (Exc./Em. 350/460 nm) [32], declining with cellular injury. Cellular samples may contain different types of autofluorescent molecules, and it is challenging to predict their distribution since they can change in time (the cell is dying or becoming apoptotic). Spectral unmixing for subtracting autofluorescence is possible using the non-negative matrix factorization variant of spectral unmixing, which exploits spectra obtained at the different excitation wavelengths [4, 33].

7   Comparison of Spectral Unmixing and Spectral Compensation

Despite extensive development in cytometry, the compensation stays based upon the classical algorithms, using the single controls approach developed by Bagwell and Adams [34], with some recent developments [35, 36]. Two methods of separating fluorophore signals in multicolor cytometry were recently compared by Niewold and colleagues [35]. One of the major limitations of spectral compensation is the increased spread of compensated signals compared to the original ones that diminish the ratio between mean/median values of positive and negative populations [37]. Particularly it precludes discrimination between negative and dim populations.

For some highly overlapping fluorophores, spectral unmixing algorithms made it possible to resolve the two fluorescence signals where spectral compensation did not. Unmixing in spectral cytometers gives less spreading, which is important when using numerous (panels >16) fluorophores [35]. However, if the cytometer uses optical filters (Aurora spectral analyzer, Cytek, USA), the quality of these filters plays a crucial role in the spreading when unmixing similar spectra [3]. The commercially available filters might slightly deviate from the characteristics provided by the supplier and, thus, sometimes, do not adequately exclude the fluorescent emission of other fluorophores and/or autofluorescent molecules that overlap with the desired signal. In commercial spectral cytometers from SONY, instead of the optical filters, specialized prism-based optics are used to measure and separate emissions from different fluorophores [38].

Another advantage of spectral unmixing in spectral cytometry is better extracting of autofluorescence signal that could be treated as an additional fluorophore [36], while compensation cannot be applied to autofluorescence until its spectrum is recorded.

8   Comparison of Spectral Cytometry and Mass Cytometry

Flow cytometry allows analysis of up to 25–40 parameters at a rate of several thousands of events per second. On the other hand, mass cytometry, currently a competitor to spectral FCM, allows typing of various immune cells on panels from 14 to 42 parameters with minimal overlap between channels and without autofluorescence [39–42]. Despite these benefits, broader practical applications of mass cytometry are affected by limitations such as slow collection rates (300–500 events/s vs. several thousand events/s. with conventional cytometry) and total cost of experimentation/ownership [43].

9   Differences and Similarities Between Spectral and Conventional Flow Cytometry

The common feature of spectral and conventional cytometry is the observation of a single cell. The full spectrum of a single event can be detected under the action of hydrodynamic focusing, where the cell passes an interrogation point and is excited by a collinear or non-collinear laser system. Subsequently, the detection of the emission signal for these two systems is fundamentally different. Spectrum detection became possible because of a unique emitting optical system. This system uses prisms and gratings to disperse fluorescence light, while a conventional cytometer splits fluorescent signal using bandpass, short pass, and long pass filters (Fig. 2). Prisms as dispersive optics in spectral FCM propagate light in a non-linear manner, unlike gratings that propagate light into a detector in a linear manner. Moreover, spectral cytometry to detect the full spectrum uses an array of detectors such as CCDs and multianode PMTs, while in most conventional configurations, separate PMT is utilized in each forward scatter (FSC), side scatter (SSC), and fluorescence channel.

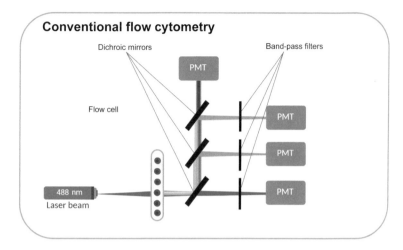

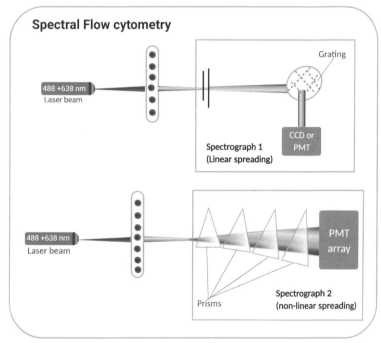

Fig. 2 The differences and similarities between spectral and conventional cytometry. Conventional cytometry: optical part – dichroic mirrors and bandpass filters. Light collection – reflection, transmission, blocking. Detectors – photomultipliers (PMT). Spectral flow cytometry: optical part – grating or prisms. Light collection - dispersion. Detectors – multianode PMTs or CCD

Further development of the real-time spectral FCM allowing measurement of emission spectra in the flow cell with the frequency typical to that of the standard flow cytometer (about 10,000 events per second) as well as the use of the spectral detectors in fluorescent microscopy was stimulated by the development of numerous fluorescent proteins with similar spectra [44]. Emission spectra of these proteins overlap significantly and thus cannot be distinguished by conventional fluorescent microscopy or FCM using dichroic mirrors and even highly selective bandpass filters [38].

This principle of spectral FCM operation is used with commercial spectral cytometry companies but with some differences in optical layout. The Sony spectral analyzer separates the emitted light with a set of prisms before sending it to 32-channel PMT arrays. To capture the fluorescence spectrum, the Cytek Aurora system employs multiple APDs with a unique set of filters in front of each APD. The possibility of obtaining a full emission spectrum with commercial spectral analyzers allowed new combinations of fluorochromes, which due to the significant spectra overlap, are not used together in conventional cytometry. Moreover, spectral FCM allows using more fluorochromes per experiment. To address the existing gap in commercially available fluorochromes, new dyes are necessary, and this need started to be addressed [45]. Another advantage of Spectral FCM is extracting the autofluorescence (AF) of cells and using it as a separate parameter(s) [46], allowing better signal resolution and even a comparison of different autofluorescent parameters [47].

10 Applications of Spectral FCM

Major problems of conventional flow cytometry can be solved using the spectral FCM: (1) enhanced number of fluorescent parameters used in a single tube (hematology, minimal residual disease (MRD)); (2) subtraction of fluorescent signal with the improvement of S/N ratio and detailed analysis of autofluorescence signal for analysis of unlabeled cells. The enhanced number of fluorescent channels is critical for analyzing small biopsies such as bone marrow aspirates in MRD. Subtraction of autofluorescence is particularly helpful for the analysis of cells with a high level of autofluorescence, such as myocytes, macrophages, brain cells, and hepatocytes. Primary cells are heterogeneous, and each subpopulation may require assigning its autofluorescence as a separate fluorophore and performing additional spectral unmixing [3].

11 Current Applications: Multi-parametric Spectral Cytometry

Nevertheless, certain studies were already made at the early stage of spectral cytometry. In 2015 Futamura and co-workers [38] described an analysis of lymphocyte migration from the individual lymph node (within 24 h) and using photoconvertible protein, and

11-color labeling showed that CD69 low naive T cell subset was replaced in lymph node faster than CD69 high memory T-cell subsets [36]. Schmutz and co-authors (2016) [48], using a two-laser Sony SP6800 instrument (405 and 488 nm), demonstrated by detailed fluorescence-minus-one control (FMO) that while the staining index (SI) for individual dyes in spectral FCM was the same as in conventional FCM, spectral FCM gives much better discrimination of dyes with similar fluorescent properties. Spectral FCM allowed discrimination of dyes with the same peak fluorescence intensity when the overall spectra were different and dyes with similar spectra but shifted for 10–20 nm peaks using Kaluza software (YFP versus GFP; both proteins versus FITC) [48].

Besides, spectral FCM allowed discrimination of lymphocytes among the cells isolated from the tissues with high autofluorescence. Complete elimination of autofluorescent signal makes it possible to discriminate dye-positive and dye-negative cells using dyes with emission spectra close to the autofluorescent spectra for further analysis [48].

The Sony SP6800 Spectral Cell Analyzer instrument utilized a 32 multianode PMT (Hamamatsu), and spectrum separation is achieved through a complex prism-based monochromator. SONY Inc. demonstrated a prototype instrument and reported on hyperspectral technology during the ISAC Congress in Seattle in 2012 and announced the launch of the new hyperspectral flow cytometer product—an SP6800 Spectral Cell Analyzer—in 2012.

In some applications, the multiplexing by spectral tags may not require spectral unmixing. In this setting, it may be beneficial to classify the spectra directly instead of classification based on unmixed intensities. Many techniques may be utilized here, including unsupervised data reduction (using, for example, principal component analysis, independent component analysis, or factor analysis) or supervised techniques (such as neural networks or support vector machines).

Advantages of spectral cytometry such as a large number of studied parameters in one panel with better resolution due to the removal of the autofluorescence signal and a rate of several thousand events per second (Sony SP6800 10,000–20,000 events/second, Cytek Aurora 35,000 events/s), have led to an increase in the practical use of spectral cytometers in immunophenotyping. One of the first multicolor panels (nine colors) was created by Futamura and co-authors [38] at the presentation of the Spectral Analyzer SP 6800 to study the movement of KikGr protein after photoconversion in the inguinal lymph node cells. The remaining immune cells, after photoconversion, changed their emission from green to red (KikGrGreen-KikGrRed) while migrated cells stayed green. In this experiment, the emission spectra of fluorochromes and fluorescent proteins, which strongly overlapped with each other, were separated using spectral unmixing (EGFP/FITC/KikGr-Green, KikGr-Red/PE, KikGr-Green/Venus, EGGP/Venus, KikGr-Red/mKO2) [36]. It would be difficult to apply this panel in conventional flow cytometry, and with the spectral analyzer, it became possible to separate and eliminate the low and high levels of autofluorescence that were found in the mouse splenocytes with strong expression of F4/80 marker (major macrophage biomarker, APC labeled) [38]. Solomon and co-authors [49] used a 15-fluorochrome panel and spectral FCM to describe the aging of the bone marrow in mice.

The separation of lasers at the Cytek spectral flow cytometer allowed the creation of 30–40 multicolor cytometric panels. The 40-color panel OMIP-069 with Aurora for identifying T cells, B cells, NKT—like cells, monocytes, and dendritic cells was reported recently [50]. This panel is effective in the study of the immune response with low sample volume [50]. In this panel, with spectral cytometry, it became possible to use dyes that have a strong overlap of the emission signal between them (PE/FITC, PE-Alexa Fluor 700/PerCP-eFluor 710, BUV 496/eFluor 450, SuperBright 436). Using data acquired by a 3-laser 38-color Aurora (Cytek, USA) spectral cytometer and analyzed by Kaluza and FlowJo software, Chen and co-authors [51] demonstrated that SFC allows distinguishing subsets of myeloid cells when using one tube with 24-color staining more precise compared to the standard 3*8-color panel. By automated clustering, malignant cells from patients with minimal residual disease (MRD) were distinguished from rare normal mast cells and basophils. In the early study, Murphy and colleagues [52] conducted a similar study for typing human peripheral blood mononuclear cells (PBMCs), but separate panels have been developed for the determination of T cells (23 colors) and B cells (22 colors). Schmutz and co-authors [48] described a 19 colors panel for the separation of murine splenocytes into B-, T-, NK-, and dendritic cells.

A new generation of SONY spectral cytometers—ID7000 also has a combination of separate lasers (for sequential excitation). It allows the use of multicolor panels, such as a 28-colors panel for immune-profiling of COVID-19 patients [53]. Two highly autofluorescent fetal liver stromal subsets were clearly discriminated using spectral unmixing with autofluorescence assigned as an independent parameter [47]. The use of other multicolor panels for immunophenotyping with a spectral cytometer is summarized in Table 1.

Table 1
Multicolor immunophenotyping panels examined by spectral FCM-type of analysis

Colors number	Instrument	Cells types	References
40	Aurora	Human PBMCs—CD4 T cells, CD8 T cells, regulatory T cells, γδ T cells, NKT-like cells, B cells, NK cells, monocytes, dendritic cells	Park et al. 2020 [54]
22–23	Aurora	Human PBMCs—T and B cells	Murphy et al., 2019 [52]
11	Sony SP 6800	KikGR expressing mice—T cells	Futamura et al., 2015 [38]
Up to 9	Sony SP 6800	Measure of CD71 expression in pDC, CD 103$^+$CD11b$^-$DC, CD103$^-$CD11b$^+$DC, AM, GR, BC, TC, NK, etc. in lung, liver, small intestine, Peyer's patches, mesenteric lymph nodes, spleen, thymus, bone marrow, blood from mouse	Lippitsch et al., 2017 [55]
12	Sony SP 6800	Mouse bladder cells—CD45, NK cells, neutrophils, macrophages, eosinophils	Rousseau et al., 2016 [56]
14	Aurora	Human PBMC—CD14, CD169 monocytes	Affandi et al., 2020 [57]
Up to 7	Aurora	Urokinase-type plasminogen activator receptor-targeted CART T cells	Amor et al., 2020 [58]
19	Sony SP 6800	Murine splenocytes—B, T, NK, dendritic, myeloid spleen cells	Schmutz et al., 2016 [48]
12	Aurora using SpectroFlo 2.2	CD8$^+$ T-cell and B-cell	Turner et al., 2020 [59]
14	Aurora using SpectroFlo	Mouse hematopoietic stem and progenitor compartments	Solomon et al., 2020 [49]
14	Aurora	Leukocytes, neutrophils, eosinophils, NK cells, NKT cells, CD4$^+$ T-cells, CD8α$^+$ T-cells, PDCs, B cells, cDC, microglia, Ly6Chigh, and Ly6Clow infiltrating monocytes	Niewold et al., 2020 [35]
12	Aurora	Composition and activation of circulating leukocytes in COVID-19 and influenza PBMCs (peripheral blood mononuclear cells (PBMCs))	Mudd et al., 2020 [60]

(continued)

Table 1
(continued)

Colors number	Instrument	Cells types	References
18	Aurora	T-cells subsets, the R-based pipeline using fluorescence minus one (FMO) controls	Fox et al., 2020 [61]
22	Aurora (3 lasers)	Splenocytes—B cells, CD4 and CD8 T-cells, neutrophils, NK cells, DCs, and monocytes. Comparison with mass cytometry	Ferrer-Font et al., 2020 [62]
9	Aurora	Placental mesenchymal stem/stromal cells	Boss et al., 2020 [63]
24	Aurora (3 lasers)	Subsets of myeloid cells for MRD analysis	Chen et al., 2020 [51]
Up to 10	Aurora (5 lasers)	Detection of murine gamma herpes virus 68 cells	Riggs et al., 2021 [46]
3 channel FRET detection	Aurora (4 lasers)	Spectral unmixing for improved FRET detection	Henderson et al., 2021 [64]
23	Aurora	23 colors for placental mesenchymal cells analysis	Boss et al., 2021 [65]
37	Aurora	45 different subpopulations, PBMC from SARS-CoV-2 infected patients	Fernandez et al., 2022 [66]
Up to 11 in one panel	ID7000	Senolytic vaccination to eliminate senescent cell in mice	Suda et al., 2021 [67]
Autofluorescence parameters	ID7000	Autofluorescence spectra analysis	Peixoto et al., 2021 [47]
9 colors for genes initially identified by RNA-sequencing	SP6800/ ID7000	CLEC12A, CD1a, CD86, CCL18, CCL17, CCL22, CD115, CD88 and CD85d	Costa et al., 2022 [68]
5 fluorescent proteins	SP6800 (SONY)	Bacterial phytochromes with far-red and near-infrared emission	[69]
Autofluorescence multiple	ID7000	autofluorescence multiple murine lung cells populations	[70]

12 Two Major Types of Spectral FCM Analysis: Virtual Filtering and Spectral Unmixing

Spectral unmixing is the most used and considered to be the most powerful approach, but it requires a thorough recording of autofluorescent controls from heterogeneous cellular populations. Sophisticated spectral unmixing with commercially available software allows robust separation from 4+ to 20+ fluorochromes. Another less powerful but more universal approach is virtual filtering. It was initially demonstrated in the phytoplankton study

[71, 72]. Spectral cytometry allowed effective selection of "filtering off" autofluorescent part of spectra, which may overlap with fluorescent signals in the multiparametric analysis of multiple taxa of algae [71]. It mimics the interchange of hardware filters in the PMT channels in a standard flow cytometer. In conventional cytometry, changing optical filters means manipulation with hardware, and some optical bandpass filters may not be available on the market. With SFC, we can make a large selection of virtual filters after the sample is recorded [72].

The use of multiple fluorescent conjugates and dyes/pigments significantly affected cytometric analysis facilitating multivariate analysis, dimensionality reduction algorithms based on stochastic neighbor embedding (SNE), unsupervised cluster analysis, and cell-subset identification programs such as SPADE, CITRUS, FlowSOM, CellCNN, and viSNE [73–77]. An alternative to clustering algorithms is principal component analysis (PCA), which is widely used in other areas of biology. Recently, Ogishi and co-authors [78] introduced iMUBAC (integration of multi-batch cytometry datasets) using unsupervised cell-type identification across multiple batches.

13 Conclusions

Currently, the spectral cytometer becomes a superior alternative to the conventional cytometer since it allows the acquisition of fluorescent dyes and proteins without the limitations of hardware optics and detectors. It leads to reducing the complexity of multi-color panel design and allows easy acquisition of more than 20 colors with good discrimination of bright, dim, and negative cellular subpopulations. The latest multi-laser (up to seven lasers) commercially available spectral cytometer ID7000 (SONY) allows the detection and analysis of up to 40 fluorescent parameters. Spectral FCM or full-spectrum cytometry can subtract autofluorescence from signals generated by dyes without increasing spread, besides, it allows acquire autofluorescence as separate parameter(s). Spectral FSM allows detailed analysis of the autofluorescence that might be especially useful for analyzing phytoplankton where a strong autofluorescent signal from chlorophyll precludes using fluorescently labeled dyes/antibodies and for highly autofluorescent cells (macrophages, myeloid progenitors, infected cells, etc.). Available libraries of emission spectra of the numerous standard fluorophores make single-stained controls unnecessary. The limitations of the spectral deconvolution approach in Spectral FCM are related to the use of tandem dyes or the inability to use ratiometric probes. The new generation of multi-laser Spectral FSM instruments initiates a breakthrough in cytometric analysis and the replacement of conventional cytometers. Full-spectrum cell sorters and co-registering spectra with images of cells can be foreseen in the near future.

Acknowledgments

Work was supported by the Ministry of Health of the Republic of Kazakhstan under the program-targeted funding of the Ageing and Healthy Lifespan research program (IRN: 51760/Ф-М P-19) and AP08857554 (Ministry of Education and Science, Kazakhstan) to IAV. NSB was funded by CRP 16482715 and SSH2020028 grants from Nazarbayev University, and AP14872088 MES grant (Kazakhstan).

References

1. Imaging flow cytometry: methods and protocols (2016) Barteneva NS, Vorobjev IA (eds) Methods in molecular biology, vol 1389. Humana, New York, 295pp

2. Wang W, Su B, Pang L et al (2020) High-dimensional immune profiling by mass cytometry revealed immunosuppression and dysfunction of immunity in COVID-19 patients. Cell Mol Immunol 17:650–652. https://doi.org/10.1038/s41423-020-0447-2

3. Maucourant C, Filipovic I, Ponzetta A, Aleman S, Cornilett M et al (2020) Natural killer cell immunophenotypes related to COVID-19 disease severity. Sci Immunol 5: eabd6832. https://doi.org/10.1126/sciimmunol.abd68

4. Rossetti BJ, Wilbert SA, Welch JLM, Borisy GG, Nagy JG (2020) Semi-blind sparse affine spectral unmixing of autofluorescence-contaminated micrographs. Bioinformatics 36:910–917. https://doi.org/10.1093/bioinformatics/btz674

5. Wade CG, Rhyne RH, Woodruff WH et al (1979) Spectra of cells in flow cytometry using a vidicon detector. J Histochem Cytochem 27:1049–1052. https://doi.org/10.1177/27.6.110874

6. Steen HB, Stokke T (1986) Fluorescence spectra of cells stained with a DNA-specific dye, measured by flow cytometry. Cytometry 7:104–106. https://doi.org/10.1002/cyto.990070117

7. Buican, TN (1990) Real-time Fourier transform spectrometry for fluorescence imaging and flow cytometry. Proceedings of the SPIE, bioimaging and two-dimensional spectroscopy 1205:126–133. https://doi.org/10.1117/12.17787

8. Gauci MR, Vesey G, Narai J et al (1996) Observation of single-cell fluorescence spectra in laser flow cytometry. Cytometry 25:388–393. https://doi.org/10.1002/(sici)1097-0320(19961201)25:4<388::aid-cyto11>3.0.co;2-r

9. Asbury CL, Esposito R, Farmer C, van den Engh G (1996) Fluorescence spectra of DNA dyes measured in a flow cytometer. Cytometry 24:234–242. https://doi.org/10.1002/(sici)1097-0320(19960701)24:3<234::aid-cyto6>3.0.co;2-h

10. Lawrence WG, Varadi G, Entine G et al (2008) Enhanced red and near infrared detection in flow cytometry using avalanche photodiodes. Cytometry A 73A:767–776. https://doi.org/10.1002/cyto.a.20595

11. Zhao S, Wu X, Chen Y, et al (2011) High gain avalanche photodiode (APD) arrays in flow cytometer opitical system. 2011 international conference on multimedia technology, pp 2151–2153. https://doi.org/10.1109/icmt.2011.6002457

12. Yamamoto M (2017) Photon detection: current status. In: Single cell analysis. Springer, Singapore, pp 227–242. https://doi.org/10.1007/978-981-10-4499-1_10

13. Nolan JP, Condello D, Duggan E, Naivar M, Novo D. (2013) Visible and near infrared fluorescence spectral cytometry. Cytometry Part A; 83A:253–264. https://doi.org/10.1002/cyto.a.22241

14. Isailovic D, Li H-W, Phillips GJ, Yeung ES (2005) High-throughput single-cell fluorescence spectroscopy. Appl Spectrosc1 59:221–226. https://doi.org/10.1366/0003702053085124

15. Robinson JP, Rajwa B, Gregori G et al (2005) Multispectral cytometry of single bio-particles using a 32-channel detector. Proc SPIE 5692: 359–365. https://doi.org/10.1117/12.591365

16. Gregori G, Patsekin V, Rajwa B, Jones J, Ragheb K, Holdman C, Robinson JP (2012) Hyperspectral cytometry at the single-cell level using a 32-channel photodetector. Cytometry A 81A:35–44. https://doi.org/10.1002/cyto.a.21120

17. Goddard G, Martin JC, Naivar M et al (2006) Single particle high resolution spectral analysis flow cytometry. Cytometry A 69A:842–851. https://doi.org/10.1002/cyto.a.20320

18. Watson DA, Brown LO, Gaskill DF et al (2008) A flow cytometer for the measurement of Raman spectra. Cytometry A 73A:119–128. https://doi.org/10.1002/cyto.a.20520

19. Nolan JP, Sebba DS (2011) Surface-enhanced Raman scattering (SERS) cytometry. Methods Cell Biol:515–532. https://doi.org/10.1016/b978-0-12-374912-3.00020-1

20. Nolan JP, Condello D, Duggan E et al (2012) Visible and near infrared fluorescence spectral flow cytometry. Cytometry A 83A:253–264. https://doi.org/10.1002/cyto.a.22241

21. Nolan JP, Condello D (2013) Spectral flow cytometry. Curr Protoc Cytom 63: 1.27.1–1.27.13. https://doi.org/10.1002/0471142956.cy0127s63

22. Sanders CK, Mourant JR (2013) Advantages of full spectrum flow cytometry. J Biomed Opt 18:037004. https://doi.org/10.1117/1.jbo.18.3.037004

23. McCausland M, Lin Y-D, Nevers T et al (2021) With great power comes great responsibility: high-dimensional spectral flow cytometry to support clinical trials. Bioanalysis 13:1597–1616. https://doi.org/10.4155/bio-2021-0201

24. de Juan A, Tauler R (2021) Multivariate curve resolution: 50 years addressing the mixture analysis problem – a review. Anal Chim Acta 1145:59–78. https://doi.org/10.1016/j.aca.2020.10.051

25. Zimmermann T, Rietdorf J, Pepperkok R (2003) Spectral imaging and its applications in live cell microscopy. FEBS Lett 546:87–92. https://doi.org/10.1016/S0014-5793(03)00521-0

26. Garini Y, Young IT, McNamara G (2006) Spectral imaging: principles and applications. Cytometry A 69A:735–747. https://doi.org/10.1002/cyto.a.20311

27. Benachir D, Deville Y, Hosseini S, Karoui MS (2020) Blind unmixing of hyperspectral remote sensing data: a new geometrical method based on a two-source sparsity constraint. Remote Sens 12:3198. https://doi.org/10.3390/rs12193198

28. Wei J, Wang X (2020) An overview on linear unmixing of hyperspectral data. Math Probl Eng 2020:1–12. https://doi.org/10.1155/2020/3735403

29. Kronick MN (1986) The use of phycobiliproteins as fluorescent labels in immunoassay. J Immunol Methods 92:1–13. https://doi.org/10.1016/0022-1759(86)90496-5

30. Monici M (2005) Cell and tissue autofluorescence research and diagnostic applications. Biotechnol Annu Rev:227–256. https://doi.org/10.1016/s1387-2656(05)11007-2

31. Croce AC, Bottiroli G (2014) Autofluorescence spectroscopy and imaging: a tool for biomedical research and diagnosis. Eur J Histochem 58:2461. https://doi.org/10.4081/ejh.2014.2461

32. Chance B (2004) Mitochondrial NADH redox state, monitoring discovery and deployment in tissue. Methods Enzymol:361–370. https://doi.org/10.1016/s0076-6879(04)85020-1

33. Rakotomanga P, Soussen C, Khairallah G, Amoroux M, Zaytsev S et al (2019) Source separation approach for the analysis of spatially resolved multiply excited autofluorescence spectra during optical clearing of ex vivo skin. Opt Express 10:3410–1424. https://doi.org/10.1364/BOE.10.003410

34. Bagwell CB, Adams EG (1993) Fluorescence spectral overlap compensation for any number of flow cytometry parameters. Ann N Y Acad Sci 677:167–184. https://doi.org/10.1111/j.1749-6632.1993.tb38775.x

35. Niewold P, Ashhurst TM, Smith AL, King NJ (2020) Evaluating spectral cytometry for immune profiling in viral disease. Cytometry A 97:1165–1179. https://doi.org/10.1002/cyto.a.24211

36. Roca CP, Burton OT, Gergelits V et al (2021) AutoSpill is a principled framework that simplifies the analysis of multichromatic flow cytometry data. Nat Commun 12:1–16. https://doi.org/10.1038/s41467-021-23126-8

37. Nguyen R, Perfetto S, Mahnke YD et al (2013) Quantifying spillover spreading for comparing instrument performance and aiding in multicolor panel design. Cytometry A 83A:306–315. https://doi.org/10.1002/cyto.a.22251

38. Futamura K, Sekino M, Hata A et al (2015) Novel full-spectral flow cytometry with multiple spectrally-adjacent fluorescent proteins and fluorochromes and visualization of in vivo cellular movement. Cytometry A 87:830–842. https://doi.org/10.1002/cyto.a.22725

39. Baumgart S, Peddinghaus A, Schulte-Wrede U et al (2016) OMIP-034: comprehensive immune phenotyping of human peripheral leukocytes by mass cytometry for monitoring immunomodulatory therapies. Cytometry A 91:34–38. https://doi.org/10.1002/cyto.a.22894

40. Jaracz-Ros A, Hémon P, Krzysiek R et al (2018) OMIP-048 MC: quantification of calcium sensors and channels expression in lymphocyte subsets by mass cytometry. Cytometry A 93:681–684. https://doi.org/10.1002/cyto.a.23504

41. Brodie TM, Tosevski V, Medová M (2018) OMIP-045: characterizing human head and neck tumors and cancer cell lines with mass cytometry. Cytometry A 93:406–410. https://doi.org/10.1002/cyto.a.23336

42. Dusoswa SA, Verhoeff J, Garcia-Vallejo JJ (2019) OMIP-054: broad immune phenotyping of innate and adaptive leukocytes in the brain, spleen, and bone marrow of an orthotopic murine glioblastoma model by mass cytometry. Cytometry A 95:422–426. https://doi.org/10.1002/cyto.a.23725

43. Iyer A, Hamers AAJ, Pillai AB (2022) CyTOF® for the masses. Front Immunol 13: 815828. https://doi.org/10.3389/fimmu.2022.815828

44. Chudakov DM, Matz MV, Lukyanov S, Lukyanov KA (2010) Fluorescent proteins and their applications in imaging living cells and tissues. Physiol Rev 90:1103–1163. https://doi.org/10.1152/physrev.00038.2009

45. Seong Y, Nguyen DX, Wu Y, Thakur A, Harding F, Nguyen TA (2022) Novel PE and APC tandems: additional near-infrared fluorochromes for use in spectral flow cytometry. Cytometry A:1–11. Epub 2 Feb 2022. https://doi.org/10.1002/cyto.a.24537

46. Riggs JR, Medina EM, Perrenoud LJ, Bonilla DL, Clambey ET, van Dyk LF, Berg LJ (2021) Optimized detection of acute MHV68 infection with a reporter system identifies large peritoneal macrophages as a dominant target of primary infection. Front Microbiol 12: 656979. https://doi.org/10.3389/fmicb.2021.656979

47. Peixoto MM, Soares-da-Silva F, Schmutz S, Mailhe M-P, Novault S, Cumano A, Ait-Mansour C (2022) Identification of fetal liver stromal subsets in spectral cytometry using the parameter autofluorescence. Cytometry A. Epub 2 May 2022. https://doi.org/10.1002/cyto.a.24567

48. Schmutz S, Valente M, Cumano A, Novault S (2016) Spectral cytometry has unique properties allowing multicolor analysis of cell suspensions isolated from solid tissues. PLoS One 11: e0159961. https://doi.org/10.1371/journal.pone.0159961

49. Solomon M, DeLay M, Reynaud D (2020) Phenotypic analysis of the mouse hematopoietic hierarchy using spectral cytometry: from stem cell subsets to early progenitor compartments. Cytometry A 97:1057–1065. https://doi.org/10.1002/cyto.a.24041

50. Park LM, Lannigan J, Jaimes MC (2020) Omip-069: forty-color full spectrum flow cytometry panel for deep immunophenotyping of major cell subsets in human peripheral blood. Cytometry A 97:1044–1051. https://doi.org/10.1002/cyto.a.24213

51. Chen M, Wang H, Fu M et al (2020) One tube 24 color full spectral flow cytometry and multi-dimensional software to study the maturation pattern and antigen expression of the myeloid. Blood 136:13–14. https://doi.org/10.1182/blood-2020-140600

52. Murphy KA, Bhamidipati K, Rubin SJS et al (2019) Immunomodulatory receptors are differentially expressed in B and T cell subsets relevant to autoimmune disease. Clin Immunol 209:108276. https://doi.org/10.1016/j.clim.2019.108276

53. Chen MC, Lai KS, Chien KL, Teng ST, Lyn YR et al (2021) Impact of placenta-derived mesenchymal stem cells treatment on patients with severe lung injury caused by COVID-19 pneumonia: clinical and immunological aspect. Res Square. Preprint. https://doi.org/10.21203/rs.3.rs-1013382/v1

54. Park LM, Lannigan J, Jaimes MC (2020) Forty-color full spectrum flow cytometry panel for deep immunophenotyping of major cell subsets in human peripheral blood. Cytometry A 97A:1044–1051. https://doi.org/10.1002/cyto.a.24213

55. Lippitsch A, Chukovetskyi Y, Baal N et al (2017) Unique high and homogenous surface expression of the transferrin receptor CD71 on murine plasmacytoid dendritic cells in different tissues. Cell Immunol 316:41–52. https://doi.org/10.1016/j.cellimm.2017.03.005

56. Rousseau M, Goh HMS, Holec S et al (2016) Bladder catheterization increases susceptibility to infection that can be prevented by prophylactic antibiotic treatment. JCI Insight 1: e88178. https://doi.org/10.1172/jci.insight.88178

57. Affandi AJ, Grabowska J, Olesek K et al (2020) Selective tumor antigen vaccine delivery to human CD169+antigen-presenting cells using ganglioside-liposomes. Proc Natl Acad Sci U S A 117:27528–27539. https://doi.org/10.1073/pnas.2006186117

58. Amor C, Feucht J, Leibold J et al (2020) Senolytic car T cells reverse senescence-associated pathologies. Nature 583:127–132. https://doi.org/10.1038/s41586-020-2403-9

59. Turner JS, Zhou JQ, Han J et al (2020) Human germinal centres engage memory and naive B cells after influenza vaccination. Nature 586:127–132. https://doi.org/10.1038/s41586-020-2711-0

60. Mudd PA, Crawford JC, Turner JS et al (2020) Distinct inflammatory profiles distinguish COVID-19 from influenza with limited contributions from cytokine storm. Sci Adv 6: eabe3024. https://doi.org/10.1126/sciadv.abe3024

61. Fox A, Dutt TS, Karger B, Rojas M, Obregon-Henao A et al (2020) CYTO-feature engineering: a pipeline for flow cytometry analysis to uncover immune populations and associations with disease. Sci Rep 10:7651. https://doi.org/10.1038/s41598-020-64516-0

62. Ferrer-Font L, Mayer JU, Old S et al (2020) High-dimensional data analysis algorithms yield comparable results for mass cytometry and spectral flow cytometry data. Cytometry A 97:824–831. https://doi.org/10.1002/cyto.a.24016

63. Boss AL, Brooks AES, Chamley LW, James JL (2020) Influence of culture media on the derivation and phenotype of fetal-derived placental mesenchymal stem/stromal cells across gestation. Placenta 101:66–74. https://doi.org/10.1016/j.placenta.2020.09.002

64. Henderson J, Havranek O, Ma MCJ, Herman V, Kupcova K, Chrbolkova T et al (2021) Detecting Förster resonance energy transfer in living cells by conventional and spectral cytometry. Cytometry A:1–17. https://doi.org/10.1002/cyto.a.24472

65. Boss AL. Damani T, Chamley LW, James JL, Brooks AES (2021) The origins of placental mesenchymal stromal: full spectrum flow cytometry reveals mesencgymal heterogeneity in first trimester placentae, and phenotypic convergence in culture. BioRxiv. Preprint. https://doi.org/10.1101/2021.12.21.473551

66. Fernandez MA, Alzayat H, Jaimes MC, Kharraz Y, Requena G, Mendez P (2022) High-dimensional immunophenotyping with 37-colors panel using full-spectrum cytometry.

67. Suda M, Shimizu I, Katsuumi G et al (2021) Senolytic vaccination improves normal and pathological age-related phenotypes and increases lifespan in progeroid mice. Nat Aging 1:1117–1126. https://doi.org/10.1038/s43587-021-00151-2

Methods Mol Biol 2386:43–60. https://doi.org/10.1007/978-1-0716-1771-7_4

68. Costa B, Becker J, Krammer T, Mulenge F, Duran V et al (2022) HCMV exploits STING signaling and counteracts IFN and ISG induction to facilitate dendritic cell infection. Res Square. Preprint, https://doi.org/10.21203/rs.3.rs-953016/v1

69. Telford WG, Shcherbakova DM, Buschke D et al (2015) Multiparametric flow cytometry using near-infrared fluorescent proteins engineered from bacterial phytochromes. PLoS One 10:e0122342. https://doi.org/10.1371/journal.pone.0122342

70. Wanner N, Barnhart J, Apostolakis N, Zlojutro V, Asosingh K (2022) Using the auto-fluorescence finder on the Sony ID7000TM spectral cell analyzer to identify and unmix multiple highly autofluorescent murine lung populations. Front Bioeng Biotechnol 10: 827987. https://doi.org/10.3389/fbioe.2022.827987. eCollection 2022

71. Dashkova V, Segev E, Malashenkov D et al (2016) Microalgal cytometric analysis in the presence of endogenous autofluorescent pigments. Algal Res 19:370–380. https://doi.org/10.1016/j.algal.2016.05.013

72. Barteneva NS, Dashkova V, Vorobjev I (2019) Probing complexity of microalgae mixtures with novel spectral flow cytometry approach and "virtual filtering". BioRxiv. Preprint. https://doi.org/10.1101/516146

73. Qiu P, Simonds EF, Bendall SC, Gibbs KD Jr, Bruggner RV et al (2011) Extracting a cellular hierarchy from high-dimensional cytometry data with SPADE. Nat Biotechnol 29:886–891. https://doi.org/10.1038/nbt.1991

74. Amir ED, Davis KL, Tadmor MD, Simonds EF, Levine JH et al (2013) viSNE enables visualization of high-dimensional single-cell data and reveals phenotypic heterogeneity of leukemia. Nat Biotechnol 31:545–552. https://doi.org/10.1038/nbt.2594

75. Van Gassen S, Callebaut B, Van Helden MJ, Lambrecht BN, Demeester P, Dhaene T, Sayes Y (2015) FlowSOM: using self-organizing maps for visualization and interpretation of cytometry data. Cytometry A 87:636–645. https://doi.org/10.1002/cyto.a.22625

76. Nowicka M, Krieg C, Weber LM, Hartmann FJ, Guglietta S et al (2017) CyTOF workflow: differential discovery in high-throughput high-dimensional cytometry datasets. F1000Res 6: 748. https://doi.org/10.12688/f1000research.11622.2

77. Bruggner RV, Bodenmiller B, Dill DL, Tibshirani RJ, Nolan GP (2014) Automated identification of stratifying signatures in cellular subpopulations. Proc Natl Acad Sci USA 111: E2770–E2777

78. Ogishi M, Yang R, Gruber C, Zhang SJ, Pelham S, Spaan AN et al (2020) Multi-batch cytometry data integration for optimal immunophenotyping. J Immunol 206:206–213. https://doi.org/10.1101/2020.07.14.202432

Chapter 2

Using Virtual Filtering Approach to Discriminate Microalgae by Spectral Flow Cytometer

Natasha S. Barteneva, Aigul Kussanova, Veronika Dashkova, Ayagoz Meirkhanova, and Ivan A. Vorobjev

Abstract

Fluorescence methods are widely used for the study of marine and freshwater phytoplankton communities. However, the identification of different microalgae populations by the analysis of autofluorescence signals remains a challenge. Addressing the issue, we developed a novel approach using the flexibility of spectral flow cytometry analysis (SFC) and generating a matrix of virtual filters (VF) which allowed thorough examination of autofluorescence spectra. Using this matrix, different spectral emission regions of algae species were analyzed, and five major algal taxa were discriminated. These results were further applied for tracing particular microalgae taxa in the complex mixtures of laboratory and environmental algal populations. An integrated analysis of single algal events combined with unique spectral emission fingerprints and light scattering parameters of microalgae can be used to differentiate major microalgal taxa. We propose a protocol for the quantitative assessment of heterogenous phytoplankton communities at the single-cell level and monitoring of phytoplankton bloom detection using a virtual filtering approach on a spectral flow cytometer (SFC-VF).

Key words Spectral flow cytometry, Phytoplankton, ID7000, Virtual filtering, Spectral flow cytometer, Cyanobacteria

---

1 Introduction

The development of spectral flow cytometry (SFC) expanded our ability to characterize heterogeneous cell populations because of the high spectral resolution achieved by this instrument [1].

The key advantage of spectral flow cytometry (SFC) is that a measurement of a set of emission spectra using different excitation wavelengths is done from individual cells with rates of hundreds and thousands of events per sec [1, 2]. Moreover, SFC analysis makes possible additional differentiation of heterogeneous algal mixtures by size and granularity in a manner similar to conventional flow cytometry (FCM) [1]. The emission spectrum information for every single cell could be combined with light scattering data

Natasha S. Barteneva and Ivan A. Vorobjev (eds.), *Spectral and Imaging Cytometry: Methods and Protocols*, Methods in Molecular Biology, vol. 2635, https://doi.org/10.1007/978-1-0716-3020-4_2, © The Author(s) 2023

through sequential gating on combinations of standard dot plots and histograms. The populations could now be separated not only by using conventional fluorescent conjugated antibodies but by also using the autofluorescent signal from unstained cells [3].

Since the spectral unmixing algorithm is based on the record of single stained probes [4], it cannot be directly applied to the natural algal probes having bright autofluorescence from different sources, and another approach has to be considered.

In 2019 we developed a novel "virtual filtering" approach (SFC-VF) based on the spectral flow cytometry analysis and use of variable regions of algal autofluorescence spectra in combination with light scattering-related separation of algal populations based on algae cellular size and granularity [5]. We applied SFC-VF to differentiate and characterize microalgae taxa in binary and multi-component mixtures as well as natural environmental microalgae assemblages and were able: (1) to differentiate microalgal cells from different phytoplankton taxa with a similar combination of pigments; and (2) to remove fluorescence signal from contaminating sources using light scatter gating. Moreover, unlike FCM, SFC makes it possible to separate individual algal cells presented in heterogenous algal populations (such as cryptophytes) based on their unique spectral data.

The SFC-VF method relies on identifying of the most variable regions of the spectra of the mixtures of algal strains analyzed pairwise and on creating a matrix of SFC fluorescent channels corresponding to those regions. Spectral differences between single algal strains (morphology—Fig. 1, left column) were captured by both spectral flow cytometer Sony SP6800 (Sony Biotechnology Inc., USA, 405 nm and 488 nm excitation) and spectrofluorimeter (Fig. 1, right column). However, the spectrofluorimeter provided an averaged signal from the population of algal cells, debris, and fluorescent organic matter. The separation of algal mixtures based on the conventional FCM approach and a filter combination used for algal analysis (such as phycoerythrin (PE) bandpass 575/25 nm) versus allophycocyanin (APC) bandpass (660/20 nm) was complicated by the heterogeneity of algal populations.

In the SFC-VF approach, the sensitivity of chlorophyll-associated channels (CH 24–30) captured on the SP6800 was switched to the minimal level. Then, the non-chlorophyll-based spectral differences (from accessory pigments) in the 420–650 nm wavelength range became prominent, enabling better discrimination of algal strains (Fig. 2). Further SFC analysis of algal cultures was continued with the reduced intensity of chlorophyll-associated channels.

Mixtures of algal cultures were analyzed in a pairwise manner generating different algal combinations. Initially, several variants of

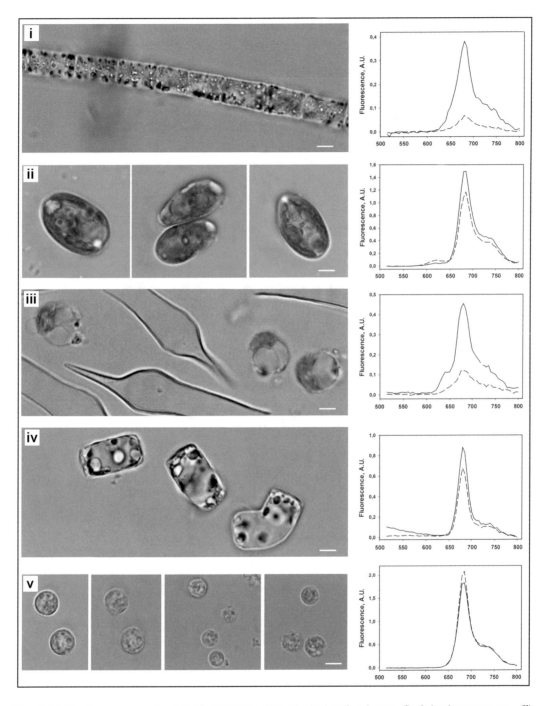

Fig. 1 Light microscopy and spectrofluorometric data of algal cell cultures. (**i**) *Aphanizomenon* sp., (**ii**) *Cryptomonas pyrenoidifera*, (**iii**) *Dinobryon divergens*, (**iv**) *Cyclotella sp.*, (**v**) *Chlorella* sp. First column: light microscopy image of algal cultures; second column: spectrofluorometric data of corresponding culture obtained with 407 nm (solid line) and 488 nm (dashed line) excitation. Scale bar 5 μm. Notice significant differences in the relative intensities at the peaks for *Aphanizomenon* sp. and *D. divergens*

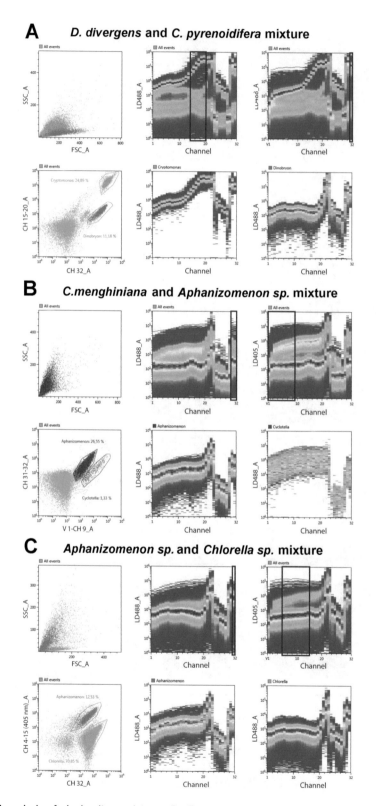

Fig. 2 Spectral analysis of algal culture mixtures *D. divergens* and *C. pyrenoidifera* (**a**), *Cyclotella sp.* and *Aphanizomenon.* (**b**), and *Aphanizomenon* and *Chlorella* sp. (**c**). Spectral data of all cells in the mixture were

a matrix of fluorescent channels corresponding to virtual filters capturing the algal spectra variability regions were created (Fig. 3).

We then selected a combination of fluorescent channels (virtual filter) that provides the best separation of two cell populations by a single dot plot. The spectra of the discriminated populations were further validated with the spectra of single algal culture controls. Furthermore, all five algal strains were mixed together and analyzed using the spectral flow cytometry analyzer. To discriminate all algal taxa, individual plot was not sufficient; instead, we used sequential gating and a combination of fluorescent channels based on virtual filters, previously selected for pairwise culture analysis (Fig. 4).

Using the above mentioned approach, we tested whether a particular microalgae type or species can be traced in the mixture of environmental microalgae populations based on its spectral profile. Different quantities (from 50% to 0.5%) of *Aphanizomenon*

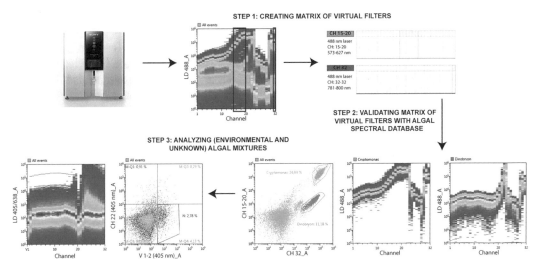

Fig. 3 Virtual filtering analysis algorithm for a mixture of microalgae cells. The mixture of microalgae cultures is analyzed using the spectral analyzer SP6800, and the obtained total spectrum of the mixture is examined for the most variable and elongated regions. A matrix of several virtual filters corresponding to the variable spectral regions is then created, and the combination of the filters providing the best separation of populations was selected (**Step 1**). The spectra of discriminated and gated populations are validated with the spectra in the algal spectral database (control spectra) (**Step 2**). In the environmental sample, virtual filters are applied, and a population different from the major one using an appropriate virtual filter could be analyzed and attributed to the cultured microalgae accordingly (**Step 3**)

Fig. 2 (continued) obtained under 488 nm laser excitation and 405 nm laser excitation spectrum charts. Based on the most variable spectral regions, combination of virtual filters corresponding to spectrum regions in channels 15–20 (488 nm excitation) and channel 32 (488 nm excitation), channels 31–32 (488 nm excitation) and channels V1-CH9 (405 nm excitation), and in channel 32 (488 nm excitation) and channels 4–15 (405 nm excitation) were selected to achieve the best discrimination of the two cell populations. Spectra of gated populations were then plotted to confirm the identity of discriminated populations

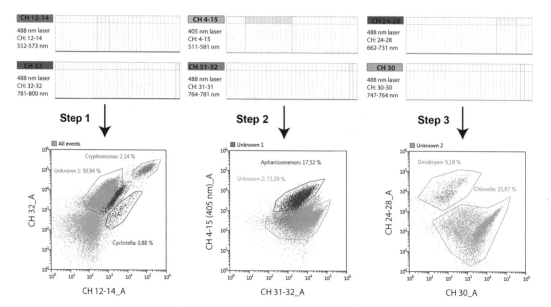

Fig. 4 Spectral analysis of five algal cultures *Aphanizomenon* sp., *C. pyrenoidifera, D. divergens, Cyclotella sp.* and *Chlorella* sp. Mixed together. *C. pyrenoidifera* and *Cyclotella sp.* populations were separated within the mixture based on CH 12–14 and CH 32 (488 nm excitation) filters (**Step 1**). Then unseparated part of the mixture (marked Unknown 1) was gated and projected onto CH 4–15 (405 nm excitation) versus CH 32 (488 nm excitation) dot plot to discriminate the cell population of *Aphanizomenon* sp. (**Step 2**). Consequently, the unidentified population (Unknown 2) was gated and visualized on a combination of CH 24–28 and CH 30 (488 nm excitation) filters to detach the last two populations of *D. divergens* and *Chlorella* sp. with very similar spectral profiles (**Step 3**)

sp. culture were mixed with environmental samples and analyzed using SFC-VF. A combination of the virtual filters CH 22 (405 nm excitation) and V1-2 (405 nm excitation) enabled the best separation of *Aphanizomenon* sp. population in the 1:1 mixture of *Aphanizomenon* sp. and environmental sample (50% *Aphanizomenon* cells: 50% pond sample) and was used for the analysis of other volume ratios. Spectra of *Aphanizomenon* sp. cells could be traced in the mixture containing as little as 0.5% proportion relative to the total volume (*see* **Note 1**).

In conventional cytometry, optical bandpass filters are used to separate fluorescent signals during instrument detection. Optimization of fluorescence detection and decreasing the acquisition of signal coming from a region with a high level of autofluorescence (e.g., GFP signal from cellular autofluorescence in a green-range region) require the replacement of a standard optical filter with a modified one [6]. The SFC-VF approach allows the creation of "virtual bandpass filters" with no hardware modification and without spectral unmixing. As a result, it was possible to narrow or widen the spectral signal that is taken into consideration from ~10 to ~300 nm bandwidth (for the SP6800 instrument) and to achieve significant discrimination of algal populations.

Initially, we analyzed representatives of five major groups of microalgae, namely (1) *Cyclotella sp.* from phylum Bacillariophyta (diatoms); (2) *Cryptomonas pyrenoidifera* from phylum Cryptista (cryptophytes; cryptomonades); (3) *Aphanizomenon* sp. from phylum Cyanobacteria ("blue-green algae", cyanoprokaryotes); (4) *Chlorella* sp. from phylum Chlorophyta ("green algae", chlorophytes); (5) *Dinobryon divergens* from phylum Ochrophyta ("golden algae"; chrysophytes) as model microalgal species with a spectral flow cytometer SP6800 (Sony Biotechnology Inc., USA). The data presented show the potential of our approach in the identification and quantitative evaluation of algal mixtures and experimental samples. In our study, we used fresh cultures; however, it is anticipated that different preservation protocols (fixation in paraformaldehyde and freezing in liquid nitrogen) may have a smoothing effect on the shape of emission spectra as it happens for the absorption spectral region related to phycobilins.

A recently introduced ID7000 instrument (Sony Biotechnology Inc., USA) is a significantly improved spectral flow cytometer compared to its predecessor Sony SP6800. It has a larger dynamic range of PMTs and an increased number of lasers (up to 7). These features make the discrimination of the algae species even simpler and more robust.

Since the dynamic range of PMTs in this cytometer is large enough, it was possible to use the standard voltage for chlorophyll channels along with other channels. The absolute amount of chlorophyll in varied species could be significantly different. Thus, to discriminate algae, a chlorophyll signal can be used.

To test the capability of ID7000 in the separation of autofluorescent spectra from different algae, we first recorded individual spectra for all three species used (Fig. 5) and denoted regions of interest there.

Next, after recording the algal mixture, we applied two regions around Chl a maximum channel (Fig. 6) representing each species and compared spectra obtained from these subpopulations with the original ones (Fig. 6d). The spectra obtained from the groups selected by these regions were nearly identical to what was measured in every single sample, proving that such selection allows good discrimination between two species.

2 Materials

2.1 Instrumentation and Accessories

1. Varioscan Flash spectral scanning multimode reader (ThermoScientific, USA).

2. The spectral flow cytometer (spectral FCM) analyzer SP6800 (Sony Biotechnology Inc., USA) was equipped with 40 mW blue 488 nm, 60 mW violet 405 nm, and 60 mW red 638 nm

Chlorella

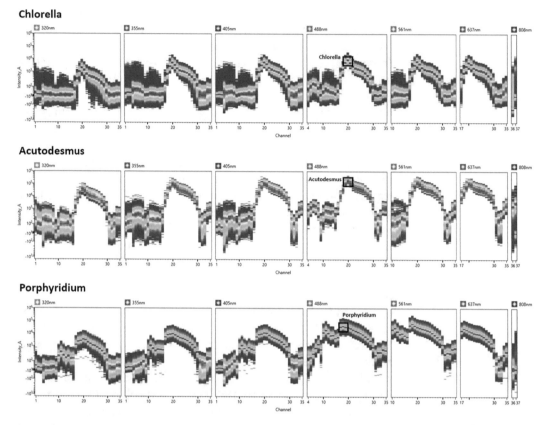

Acutodesmus

Porphyridium

Fig. 5 Spectral analysis of algal cultures *Chlorella* sp., *Acutodesmus obliquus, Porphyridium sordidum* with ID7000 spectral flow cytometer (Sony Biotechnology Inc., USA). For further analysis, regions of interest (ROI) were created in the Chl a channels (shown in black in 488 nm spectra)

lasers and a 32-channel linear array photomultiplier (500–800 nm range for 488 nm excitation and 420–800 nm range for 405/638 lasers combination), and acquisition and analysis software

3. The spectral flow cytometer (spectral FCM) analyzer ID7000 (Sony Biotechnology Inc., USA) was equipped with 20 mW deep UV 320 nm, 50 mW UV 355 nm, 100 mW violet 405 nm, 150 mW blue 488, 100 mW yellow-green 561 nm, 140 mW red 637 nm lasers and 150 mW far red 808 nm, 186 detectors: 184 fluorescence channels, one forward scatter, one side scatter, and equipped with ID7000 acquisition and analysis software (Sony Biotechnology Inc., USA).

4. Algae growth and harvesting chamber Percival model AL-30L2 (Percival Scientific Inc., USA) for algal culture incubation (with controlled temperature, light, and humidity conditions).

5. Brightfield microscope Axiovert with a color camera (Carl Zeiss Inc., Germany) (*see* **Note 2**).

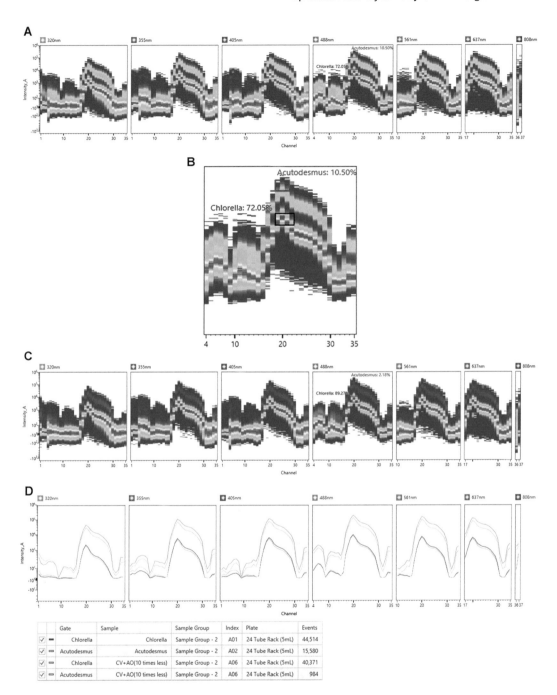

Fig. 6 Spectral analysis of mixed algal cultures *Chlorella* sp. and *Acutodesmus obliquus*, with ID7000 spectral flow cytometer. (**a**) All spectra, mixture in a ratio 1:7. (**b**) Enlargement of the spectra excited from 488 nm laser. ROI used for the selection of each species are shown as black rectangles. (**c**) All spectra, algal mixture in a ratio 1:50. The same ROI were applied for the species selection. (**d**) Comparison of the spectra obtained from pure samples and by selection using ROI

2.2 List of Microalgae Cell Cultures

Microalgae cell cultures from major microalgae taxa, including *Cyclotella sp.* CCMP334, *Chlorella* sp. CCMP251, *Dinobryon divergens* CCMP3055, *Cryptomonas pyrenoidifera* CCMP1177, *Aphanizomenon* sp. CCMP2764, *Acutodesmus obliquus* SAG 276-1, and *Porphyridium sordidum* SAG 114.79 were obtained from the National Center for Marine Algae and Microbiota (NCMA; Bigelow Laboratory for Ocean Sciences, USA) and Göttingen University's collection of algal cultures (Germany).

2.3 Reagents

1. Microalgae cell culture media: (1) DY-V medium; (2) L1 medium; (3) L1 derivative, L1–11 psu medium.

2. Eight peak beads (Sony Biotechnology Inc., USA).

3. Align Check beads (Sony Biotechnology Inc., USA).

4. 12 × 75 mm round-bottom Falcon polystyrene tubes.

3 Methods

3.1 Cultivation of Algae Cultures

Freshwater cultures *D. divergens*, *Aphanizomenon* sp., and *C. pyrenoidifera* were maintained in DY-V medium (modified from Lehman and co-authors [7]) at 14 °C and 20 °C, respectively, under 150 μmoles/m^2/s light irradiance and 12/12 L/D cycle. *Chlorella* sp. and *Cyclotella sp* were maintained in L1 medium and L1 derivative, L1–11 psu medium, respectively, at 14 °C under 150 μmoles/m^2/s light irradiance and 12/12 L/D cycle. 1. Two or more phytoplankton cell cultures (e.g., *Chlorella* sp. CCMP1177, *Acutodesmus obliquus* SAG-276-1 and *Porphyridium sordidum* SAG 114.79) were used for experiments with ID7000 spectral flow cytometer.

3.2 Spectrocytofluorimetric Acquisition of Microalgal Samples

1. Prior to the analysis, spin down each microalgae culture and resuspend it in a small volume. Count algal cells (microscope). For spectral cytometry analysis, 1000 μL volume of each culture should be used to analyze single culture controls.

2. Spectral analysis of algal cell cultures for ID 6800:

3. Prepare mixtures using 500 μL volume of each culture to analyze 10 pairwise culture mixtures and 200 μL volume of each culture to analyze a mixture of all five cultures together (ratio 1:1:1:1:1). Alternatively, mix *Chlorella* sp. and *Acutodesmus obliquus* cultures with relatively equal cell densities in 1:1, 1:10 volume ratio making up to 250 μL sample. The next steps are accommodated for the use of the ID7000 spectral system (Sony Biotechnlogy Inc., USA).

4. Turn on the spectral flow cytometer and run autocalibration using calibration beads. Use Ultra Rainbow calibration beads (Spherotech, USA or Sony Biotechnology Inc., USA) for automatic calibration.

5. Open the Acquisition window in the ID7000 software.

6. Prime the fluidics lines by flushing with sheath fluid.

7. Dissolve two drops of Align Check beads in 450 µL water.

8. Dissolve two drops of eight-peak beads in 450 µL of water.

9. Run the Daily and Performance QC.

10. Spectral analysis of single algal cell cultures. For SP6800 spectral cytometer: Adjust the laser power for 488 nm and for 405 nm lasers; reduce gain for channels 24–32 to the minimum and adjust gain for other channels. Record emission spectra of single cells in the range 420–800 nm using excitation at 405/407 nm and in the range 500–800 nm using excitation at 488 nm for SP6800.

11. Record mixed samples.

3.3 Spectral Analysis of Algal Cell Cultures for ID 7000

1. Choose Template—24 Tube Rack in the Experiment tab.

2. Load adjusted for different microalgae ID7000 settings (FSC to—16, SSC gain to 30, the threshold value to 11%, and fluorescence PMT voltage from 40% to 70%).

3. Set the sample flow rate to 1 under the "Flow Control" tab to keep the intermediate flow velocity.

4. Set the stopping condition to 50,000.

5. Create FSC_A vs. SSC_A dot plot and ribbon plot for all lasers. (*see* **Note 3**).

6. Place the round-bottom tube with the *Chlorella* sp., *Acutodesmus obliquus,* and two mixed cultures at different ratios in 24 Tube rack.

7. Place the rack in the multi-well plate holder and click "Load".

8. Highlight sample positions as a "Target" and move all samples to sample group 1.

9. Choose Set current position in the first sample tube by right-clicking, and then click "Preview."

10. Once the sample is being processed, observe if any parameters from the "Detector & Threshold," e.g., fluorescence PMT voltage and/or FSC/SSC gain, need to be tuned.

11. After tuning, click "Auto Acquire" to record the samples (*see* **Notes 4–7**, Fig. 5).

12. Record mixed samples with ID 7000 spectral flow cytometer (Fig. 6).

3.4 Analysis of Microalgal Samples on a Spectral Flow Cytometer (Example with Cultures Acquired with ID7000 Spectral Flow Cytometer)

1. Once the acquisition is completed, go to the Analysis window and open the recorded experiment.

2. Select the tube corresponding to the *Chlorella* sp., and on the FSC/SSC plot, place the gate to exclude debris and by right-clicking on the plot area, choose Sync Scale and Gate.

3. Place the gate on a 488 nm ribbon plot to include the dense events (ap. CH 18–22, 657–712 nm), name the gate "Chlorella", and create a ribbon plot from "Chlorella" gate.

4. Repeat the previous step for *Acutodesmus obliquus* culture.

5. Since all gating is synchronized within one group, go to the *Chlorella* sp. sample and send the "Chlorella" ribbon plot for "Overlay" by right-clicking.

6. Send the "Chlorella" ribbon plot to "Overlay" from the mixed sample.

7. Repeat steps previous steps for *Acutodesmus obliquus* culture.

8. Observe the differences and similarities between spectra in "Overlay" builder.

3.5 Virtual Filtering (Example with Cultures Acquired with ID6800 Spectral Flow Cytometer)

The virtual filtering algorithm is shown in Figs. 3 and 4 and consists of the following steps:

1. Using spectra from individual algal cultures, select two channels (or groups of channels) with a maximal difference in the intensity of the signal between two cultures when using settings determined in the steps above (*see* **Note 8**).

2. Make dot plots using two virtual channels along the X and Y axes for each pair, the mixture, and the environmental sample. The details are shown in Fig. 3.

3. Change virtual filters by adding or removing channels to achieve maximal separation for two populations. The details are shown in Fig. 4. (*see* **Notes 9–10**).

These filters will be further used for PCA (Principal Component Analysis) (Fig. 7) (*see* **Notes 10** and **11**).

3.6 Analysis of Microalgal Environmental Populations

The analysis illustrated an example where *Aphanizomenon* sp. was determined.

1. Record your sample. Analyze it with the virtual filters prepared for mixed strains and try to select the side population (as shown in Fig. 8, upper row).

2. If this side population is similar to one of your strains, then make a mixture of this strain with an environmental sample in 1:1 and 1:10 proportions and run this sample. Your suggestion is correct if this side population in the mixed sample will be enlarged and its position will not be shifted (**Note 11**).

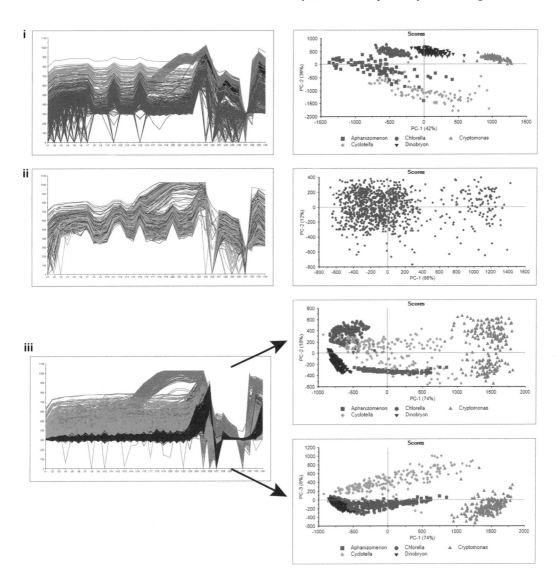

Fig. 7 Principal component analysis (PCA) performed for spectral data of algal cultures *Aphanizomenon* sp., *C. pyrenoidifera*, *D. divergens*, *Cyclotella sp.* and *Chlorella* sp. (**i**) Projection of spectra of individual cells (left) of artificially mixed algal cultures onto the plane of the first two principal components (PC) (right). (**ii**) Projection of spectra of individual cells (left) of physical mixture of algal cultures onto the plane of the first two PCs (right). (**iii**) Projection of spectra of individual cells (left) of FCM gated populations from the mixture of algal cultures onto the plane of the PC1 and PC2 (top right) and the plane of the PC1 and PC3 (bottom right). Four populations are clearly discriminated on the PC1/PC2 dot plot. The population of *Cyclotella* appears to be heterogeneous. On the PC2/PC3 plot, *Cryptomonas* and *Cyclotella* populations are more compact, but *Chlorella* population cannot be discriminated from *Aphanizomenon* sp. and *Dinobryon divergens*

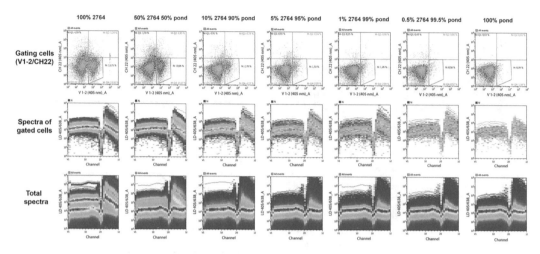

Fig. 8 Tracing different quantities of CCMP2764 *Aphanizomenon* sp. cells in an environmental sample from a pond based on spectral characteristics. From left to right: 100% of 2764 cell culture, 50% volume of 2764 culture and 50% volume of pond sample, 10% volume of 2764 culture and 90% volume of pond sample, 5% volume of 2764 culture and 95% volume of pond sample, 1% volume of 2764 culture and 99% volume of pond sample, 0.5% volume of 2764 culture and 95.5% volume of pond sample, and 100% of pond sample. In the first row: all cells are displayed on channel 22 (405 nm laser excitation) versus channels V1–2 (405 nm laser excitation) density plot, and a region corresponding to 2764 cells region is gated (L). Second row: spectra of gated L regions are displayed on 405 nm/638 nm spectrum plots. Third row: all cells in the sample are displayed on 405 nm/638 nm spectrum plots

3.7 Auto-fluorescence Finder

Algae samples can also be analyzed with an Autofluorescence Finder Tool. The ID7000 software allows to consider autofluorescence as an independent fluorescent parameter, which is particularly important for the taxonomic identification of different algae. Adding the 320 nm laser allows to include in the algal fluorescent signature a unique emission region far from Chl peak, and highly variable for different algae.

The part of ID7000 analytical software—Autofluorescence Finder Tool allows detecting the independent autofluorescent signals (Fig. 6). After finding the best separation between autofluorescent populations, the events are gated, and their emission spectra are checked. Each population should have its unique autofluorescence spectrum (Fig. 9). Autofluorescence Finder Tool in the analysis of algae also helps to separate dead algae from living ones due to the difference in the autofluorescence. The example of *Chlorella* shows three distinct algae populations with unique autofluorescent spectra (AF-A dead, AF-B aggregates, AF-C live *Chlorella*) acquired from seven lasers, including a new 320 nm UV-laser. Dead and aggregated algae can be cut off (Fig. 10), and live populations of algae with a more evident spectrum can be used in further analysis (*see* **Note 12**).

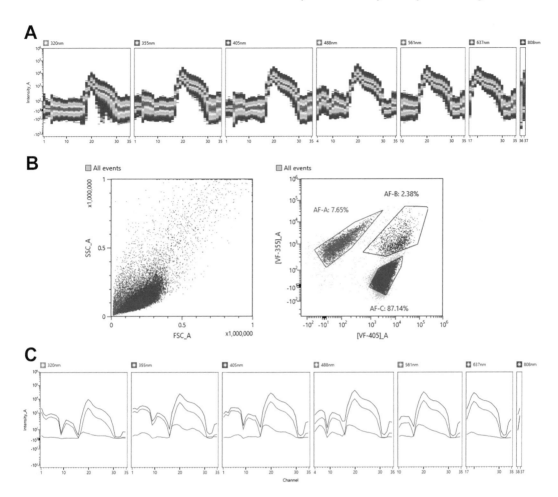

Fig. 9 *Chlorella* sp. analysis using Autofluorescence Finder Tool. (**a**) Ribbon spectral plots for the entire sample; (**b**) Gating by Autofluorescence Finder Tool using 355 nm and 405 nm excitation lasers; (**c**) Three autofluorescence spectra for the selected subpopulations (red: debris and dead cells; blue: large cells with high Chl a level; violet: cells losing Chl a photosynthetic activity)

4 Notes

1. Notably, a small population of cells with a spectral profile similar to *Aphanizomenon* sp. was detected in the gated region of an environmental sample, which can be explained by the presence of similar or same cyanobacteria species in the collected sample.

2. Any type of brightfield microscope equipped with a high NA objective (i.e., 60× or 100× with NA1.3-1.4) and a good enough color camera can be used.

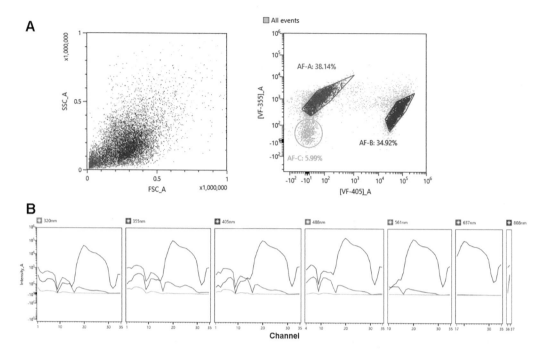

Fig. 10 *Acutodesmus obliquus* analysis using Autofluorescence Finder Tool. **(a)** Gating by Autofluorescence Finder Tool using 355 nm and 405 nm excitation lasers; **(b)** Three autofluorescence *Acutodesmus*. spectra – of debris (green); dying (red) cells losing Chl a autofluorescence, and alive (blue) algal cells

3. Since the maximal difference between algae is determined for the accessory photosynthetic proteins, the maximal gain should be used for the channels with minimal chlorophyll emission (CH 10–24 representing wavelengths 500–630 nm).

4. In the photosynthetic algae or cyanobacteria population, dead cells could be easily removed from the sample using a relatively low Chl a signal.

5. The residual difference between spectra obtained in the single probe and from the region while in a mixture, could be explained by a limited region while the whole distribution used for calculating the mean in the last figure is broader.

6. When recording all samples, you cannot change the threshold and the parameters in a detector.

7. The optimal cell concentration of microalgae cultures is in the 20,000–75,000 cells mL^{-1} range. A higher concentration will result in a high frequency of doublets. Record not less than 50,000 events in each probe using the settings.

8. Virtual filters for the dot plot data could be generated using Sony software v1.6 (Sony Biotechnology Inc., USA) and FlowJo software vs.10.2 (Treestar, USA).

9. We used CH 32 or CH 31 as one virtual filter and different groups of channels for another virtual filter (SP 6800, Sony Biotechnology Inc., USA). The best separation between pairs of algal cultures is achieved using different filter settings.

10. Some of the algal populations (*Chlorella* sp. and *Acutodesmus obliquus*. in our samples) might have a remarkably high number of cells with 0 values in channels 24–27, which may be associated with low chlorophyll signal due to the death of the cells. In order to reduce the cell heterogeneity within the sample, cells with no chlorophyll signal should not be included in the statistical analysis. Only gated chlorophyll-positive populations are shown in Fig. 5 (two upper rows). However, such gating cannot be applied to algae with a low level of chlorophyll.

11. Analysis of the environmental population will be successful only in determining the presence of one of the cultures that have been tested in advance.

12. We used a single step for separation since our algae were different in Chl a fluorescent intensity. However, when median Chl a intensities between compared species are similar, other channels should be taken into account.

Acknowledgements

This work was supported by CRP grant 16482715 (Nazarbayev University) and AP14872028 (Ministry of Education and Science, Kazakhstan) to N.S.B.; grants AP14869915 (Ministry of Education and Science, Kazakhstan) and 240919FD3937 (Nazarbayev University) to I.A.V.

References

1. Gregori G, Patsekin V, Rajwa B, Jones J, Ragheb K, Holdman C, Robinson JP (2012) Hyperspectral cytometry at the single-cell level using a 32-channel photodetector. Cytometry A 81:35–44. https://doi.org/10.1002/cyto.a.21120

2. Sanders CK, Mourant JR (2013) Advantages of full spectrum flow cytometry. J Biomed Opt 18:037004. https://doi.org/10.1117/1.JBO.18.3.037004

3. Peixoto MM, Soares-da-Silva F, Schmutz S, Mailhe M-P, Novault S, Cumano A, Ait-Mansour C (2022) Identification of fetal liver stroma in spectral cytometry using the parameter autofluorescence. Cytometry A 2022:960. https://doi.org/10.1002/cyto.a.24567

4. Futamura K, Sekino M, Hata A et al (2015) Novel full-spectral flow cytometry with multiple spectrally adjacent fluorescent proteins and fluorochromes and visualization of in vivo cellular movement. Cytometry A 87:830–842. https://doi.org/10.1002/cyto.a.22725

5. Barteneva NS, Dashkova V, Vorobjev I (2019) Probing complexity of microalgae mixtures with novel spectral flow cytometry approach and "virtual filtering". BioRxiv:516146. https://doi.org/10.1101/516146

6. Vorobjev IA, Buchholz K, Prabhat P, Ketman K, Egan ES, Marti M, Duraisingh MT, Barteneva NS (2012) Optimization of flow cytometric detection and cell sorting of transgenic Plasmodium parasites using interchangeable optical filters. Malar J 11:312. https://doi.org/10.1186/1475-2875-11-312

7. Lehman JT (1976) Ecological and nutritional studies on Dinobryon Ehrenb.: seasonal periodicity and the phosphate toxicity problem. Limnol Oceanogr 21:646–658. https://doi.org/10.4319/lo.1976.21.5.0646

Part II

Imaging Flow Cytometry: Imagestream Systems

Chapter 3

Imaging Flow Cytometric Analysis of Primary Bone Marrow Erythroblastic Islands

Joshua Tay, Kavita Bisht, Ingrid G. Winkler, and Jean-Pierre Levesque

Abstract

The erythroblastic island (EBI) is a multicellular functional erythropoietic unit comprising a central macrophage nurturing a rosette of maturing erythroblasts. Since the discovery of EBIs more than half a century ago, EBIs are still studied by traditional microscopy methods after enrichment by sedimentation. These isolation methods are not quantitative and do not enable precise quantification of EBI numbers or frequency in the bone marrow or spleen tissues. Conventional flow cytometric methods have enabled quantification of cell aggregates co-expressing macrophage and erythroblast markers; however, it is unknown whether these aggregates contain EBIs as these aggregates cannot be visually assessed for EBI content. Combining the strengths of both microscopy and flow cytometry methods, in this chapter we describe an imaging flow cytometry method to analyze and quantitatively measure EBIs from the mouse bone marrow. This method is adaptable to other tissues such as the spleen or to other species provided that fluorescent antibodies specific to macrophages and erythroblasts are available.

Key words Erythroblastic island, Macrophage, Imaging flow cytometry, Erythropoiesis

1 Introduction

In conventional flow cytometry, events are gated and quantitively analyzed according to their optical properties (such as forward scatter or side scatter of a laser beam at a specific wavelength), fluorescence intensity once the sample has been incubated with fluorescent probes or fluorescent antibodies specific for antigens present at the surface or inside cells, etc. The advantage of conventional flow cytometry is the high throughput (from thousands to millions of events in a few minutes) with simultaneous detection and quantification of an increasing number of optical and fluorescence parameters. For instance, the most advanced high-end flow cytometers, such as the BD FACSymphony A5 equipped with 6 different excitation lasers, can detect up to 47 distinct fluorescent probes/antibodies simultaneously. However, despite these phenomenal technical advances, conventional flow cytometers are

Natasha S. Barteneva and Ivan A. Vorobjev (eds.), *Spectral and Imaging Cytometry: Methods and Protocols*,
Methods in Molecular Biology, vol. 2635, https://doi.org/10.1007/978-1-0716-3020-4_3,
© Springer Science+Business Media, LLC, part of Springer Nature 2023

blind in the way that they do not allow simultaneous visualization of the analyzed events.

In-flight imaging flow cytometry (IFC) enables the high throughput direct visualization of "events" that are routinely detected by conventional flow cytometry. This is particularly advantageous when the subcellular location of particular proteins, ions, or phosphorylation event is to be measured and quantified on a large number of cells in a relatively short period of time. Consequently, IFC has been particularly useful to detect, for instance, receptor clustering at the cell surface [1], protein phosphorylation [2], subcellular localization of transcription factors [3] or organelles [4], or calcium flux [5] following cell stimulation.

Another area where IFC is advantageous is the study of biologically relevant functional multicellular units. The bone marrow of vertebrates is a rich source of such functional multicellular domains called "niches." Niches are formed by specific microenvironmental arrangements of particular cell types, extracellular matrix, and physical conditions in order to control and regulate the behavior of hematopoietic stem and progenitor cells [6, 7]. While hematopoietic stem cell niches have not been studied by IFC yet, IFC is proving very useful to better characterize cell aggregates that form erythropoietic niches in the bone marrow called erythroblastic islands [8, 9].

Erythroblastic islands (EBIs) were discovered in the late 1950s as multicellular units formed by a central macrophage rosetted by erythroblasts at various stages of erythroid differentiation [10]. EBIs are key functional units in which the central macrophage provides growth factors, iron for hemoglobin synthesis, and, very importantly, facilitates the enucleation of mature erythroblasts necessary to form mature enucleated erythrocytes [11, 12]. For decades, EBIs were studied by microscopy after long cell preparation from the bone marrow or spleen, and the fact that these cells preparations are a non-homogeneous mixture of single cells, non-specific cell aggregates, and EBIs makes any quantification of EBI frequency or number in a given tissue or following a treatment very difficult. Application of IFC to EBIs has enabled better characterization of the phenotype and functional characteristics of EBIs [8, 13] and high throughput quantification of their numbers and frequencies in the bone marrow in response to treatments [9, 14]. This will enable the study of EBIs under different external stressors (e.g., inflammation, treatments, etc.) that cause reduced erythropoiesis and anemia [9, 14].

In this chapter, we describe a method for the quantification and characterization of uncultured bone marrow murine EBIs utilizing 7 cell surface antigens in addition to the nucleus and morphological parameters on the advanced 12-channel Amnis ImageStream^X Mk II imaging flow cytometer. This technology is well equipped to objectively quantify the relatively rare, fragile, and morphologically varied EBI population as it combines the capability to filter through thousands of aggregates quickly using flow cytometry and present morphometric and localization of immunophenotypic markers using microscopy. Although this method describes the quantification of EBIs from flushed bone marrow samples, this method can be easily extrapolated to study EBIs and erythroblast maturation from other organs or other species.

2 Materials

2.1 Collection of Bone Marrow

1. 7–9 week old C57BL/6 J mice (see **Note 1**).
2. EBI buffer: Iscove's Modified Dulbecco's Medium, 20% fetal bovine serum (Gibco), 100 U/mL penicillin, 100 μg/mL streptomycin, 2 mM L-glutamine (see **Note 2**).
3. 1 mL syringes attached to 23G needles.
4. Single-edged scalpel blade on blade holder.
5. Dissecting scissors and forceps.
6. 4 mL round bottom tubes.
7. 70 μm nylon cell strainer fitted on a 50 mL conical tube (see **Note 3**).
8. Hematology analyzer or hemocytometer and microscope.

2.2 Staining and Preparation of EBIs

1. 1.5 mL microfuge tubes.
2. Swinging bucket centrifuge.
3. Freshly prepared 4% paraformaldehyde (PFA) dissolved in phosphate-buffered saline pH = 7.4 (PBS).
4. PBS NCS: Phosphate-buffered saline, 2% newborn calf serum (Gibco).
5. Blocking buffer: culture supernatant from 2.4G2 hybridoma producing neutralizing rat anti-mouse CD16/CD32 produced in-house (see **Note 4**).
6. Anti-mouse antibodies (all from Biolegend) (see **Note 5**):
   CD169-fluorescein isothiocyanate (FITC) (clone 3D6.112).
   CD71-phycoerythrin (PE) (clone R17217).
   CD11b-PE-CF594 (clone M1/70).
   Anti-Ter119-peridinin-chlorophyll-cyanine 5.5 (PerCP-Cy5.5) (clone Ter119).
   Anti-VCAM1-PE-Cy7 (clone 429 [MVCAM.A]).
   Anti-F4/80-allophycocyanin (APC) (clone BM8).
   Anti-Ly6G-APC-Cy7 (clone 1A8).
7. Hoechst33342 (Invitrogen).
8. Orbital shaker/incubator (optional).

2.3 IFC Sample Acquisition and Analysis

1. Amnis ImageStream^X Mk II imaging flow cytometer (*see* **Note 6**).

2. IDEAS analysis software (Amnis Corporation).

3   Methods

3.1 Isolation of Bone Marrow

1. Euthanize mouse (*see* **Note 7**).

2. Remove hind limb. Dissect and clean femur.

3. Open marrow cavity on both ends with a scalpel just distal to the femoral head and greater trochanter and immediately proximal to the patellar groove.

4. Gently flush marrow out in a proximal to distal direction (hip to knee) with 1 mL ice-cold EBI buffer in 1 mL syringe with 23G needle by placing needle into the epiphysis of the femur.

5. Gently flush residual marrow out in a distal to proximal direction (knee to hip) with another 1 mL ice-cold EBI buffer in the same way as **step 4** (*see* **Note 8**).

6. Triturate marrow by drawing suspension in and out of a 1 mL pipette tip fitted on pipette (*see* **Note 9**).

7. Pass cell aggregate suspension through 70 μm cell strainer into a 50 mL tube and keep samples on ice.

8. Count single cells in suspension by hematology analyzer (or hemocytometer) and record number of white blood cells per ml of cell suspension, which will be used to determine total EBIs per femur following IFC analysis.

3.2 Fixation, Staining, and Preparation for IFC

1. Centrifuge 10^7 bone marrow cells in a 1.5 mL microfuge tube at 200 relative centrifugation force (RCF) for 5 min at 4 °C.

2. Aspirate all but 20 μL of the cell pellet supernatant and resuspend pellet gently.

3. Add 500 mL of freshly made PFA fixative to all samples and single-color controls and incubate at room temperature for 10 min (*see* **Note 10**).

4. Wash fixed cell aggregates with 1 mL PBS NCS and centrifuge for 5 min at 200 RCF at room temperature for 5 min.

5. Aspirate all but 20 μL of the cell pellet supernatant and resuspend pellet gently.

6. Repeat **steps 4–5** to remove residual PFA.

7. Add 50 μL of EBI stain cocktail (1:50 CD169-FITC, 1:100 CD71-PE, 1:100 CD11b-PE-CF594, 1:40 anti-Ter119-PerCP-Cy5.5, 1:50 anti-VCAM1-PE-Cy7, 1:150 anti-F4/80-APC, 1:100 anti-Ly6G-APC-Cy7, and 5 μM Hoechst33342, all final concentrations) dissolved in blocking

buffer to all samples (*see* **Note 11**). Use the same antibody at the same dilution for single-stain color controls.

8. Incubate on an orbital shaker at 37 °C for 1 h in the dark.

9. Wash cell aggregates with 1 ml PBS NCS and centrifuge for 5 min at 200 RCF at room temperature.

10. Remove all but ~20 μL of supernatant and gently resuspend pellet (*see* **Note 12**).

11. Keep samples in the dark until sample acquisition.

3.3 IFC Setup and Acquisition

1. Channel (Ch) setup on 2-camera ImageStream^X Mk II (*see* **Notes 13–16**):

   Ch01 (457/45 nm): Brightfield 1.

   Ch02 (527/65 nm): CD169-FITC.

   Ch03 (577/35 nm): CD71-PE.

   Ch04 (610/30 nm): CD11b-PE-CF594.

   Ch05 (702/85 nm): anti-Ter119-PerCP-Cy5.5.

   Ch06 (762/35 nm): anti-VCAM1-PE-Cy7.

   Ch07 (457/45 nm): Hoechst33342.

   Ch09 (577/35 nm): Brightfield 2.

   Ch11 (702/85 nm): anti-F4/80-APC.

   Ch12 (762/35 nm): anti-Ly6G-APC-Cy7.

2. Additional settings:

   Excitation lasers: 405 nm laser at 10 mW, 488 nm laser at 100 mW, and 642 nm laser at 150 mW (*see* **Note 17**).

   60× objective lens and low acquisition speed (*see* **Note 18**).

   Singlet gate: Set Area of Brightfield 1 (Ch01) at 50–200 μm^2 and Aspect ratio of Brightfield 1 (channel 1) at 0.8–1 (*see* **Note 19**).

   Aggregate gate: Set Area of Brightfield 1 (Ch01) 300–2500 μm^2 (*see* **Note 20**).

   All cells gate: Set Area of Brightfield 1 (Ch01) 50–2500 μm^2 (*see* **Note 21**).

3. Start running sheath fluid without sample loaded (*see* **Note 22**).

4. Load the first EBI panel stained sample with the brightest expected fluorescence intensities in all channels of interest (*see* **Note 23**).

5. Record total sample volume shown on top left of the INSPIRE software interface.

6. Ensure max pixel intensity does not exceed 4095 (saturation of signal) in each channel of interest and adjust laser power as required.

7. Acquire and save 10,000 events in the "all cells" classifier (*see* **Note 24**).

8. Acquire and save 10,000–30,000 events in the "cell aggregates" classifier files and record residual sample volume (*see* **Notes 25** and **26**).

9. Load the next sample and repeat **steps 5–8** for all samples.

10. Collect single-stained color controls using the singlet cell classifier for spectral compensation (*see* **Note 27**).

3.4 IFC Gating Strategy for Gated Erythroid–Macrophage Aggregates (GEMAs)

1. Perform spectral compensation using single-stained color controls in the IDEAS software to obtain a compensation matrix file (.ctm) (*see* **Note 28**).

2. Apply compensation matrix to a sample raw image file (.rif) to create a data analysis file (.daf) (*see* **Note 29**).

3. Create "New Scatterplot" from total events with Aspect ratio Brightfield 1 vs Area Brightfield 1 and gate on "aggregates" as mentioned in **step 2** of Subheading 3.3 and shown in Fig. 1a (*see* **Note 29**).

4. Create "New Scatterplot" from the "aggregates" gate with Area Brightfield 1 vs Ch07 intensity (Hoechst33342) and gate on "Hoechst33342$^+$" events as in Fig. 1b.

5. Create "New Scatterplot" from the "Hoechst33342$^+$" gate with Ch11 intensity (F4/80-APC) vs Ch05 intensity (Ter119-PerCP-Cy5.5) and gate on "F4/80$^+$ Ter119$^+$" events as in Fig. 1c (*see* **Note 30**).

6. Create "New Scatterplot" from "F4/80$^+$ Ter119$^+$" gate with Ch03 intensity (CD71-PE) vs Ch02 intensity (CD169-FITC) and gate on "CD169$^+$ CD71$^+$" events as in Fig. 1d.

7. Create "New Scatterplot" from "CD169$^+$ CD71$^+$" gate with Ch06 intensity (VCAM1-PE-Cy7) vs Area Brightfield 1 and gate on VCAM1$^+$ events as in Fig. 1e (called gated erythroid–macrophage aggregates or GEMAs) [9].

3.5 IFC Manual Screening of GEMAs for Potential EBIs

1. Click on the "Image Properties" button and the "Display Properties" tab. Choose the desired color for each marker. Adjust the "Image Display Mapping" pixel intensity thresholds for each channel to optimize viewing (*see* **Notes 31** and **32**).

2. Click on the "Image Properties" button and the "Composite" tab. Click "New", then "Add Image" and select "Ch01" (Brightfield 1) at 10%, "Ch05" (Ter119-PerCP-Cy5.5) at 100%, "Ch07" (Hoechst33342) at 100%, and "Ch11" (F4/80-APC) at 100% to create the new composite image

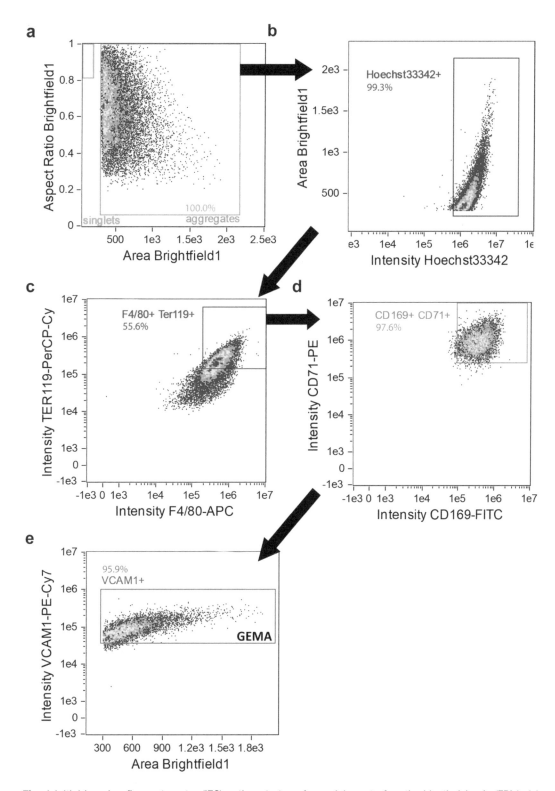

Fig. 1 Initial imaging flow cytometry (IFC) gating strategy for enrichment of erythroblastic islands (EBIs). (**a**) Only aggregates with Area of Brightfield 1 between 300–2500 μm^2 were acquired for analysis. (**b**–**e**) EBIs

"Ch01/Ch05/Ch07/Ch11" (*see* **Note 32**). This composite image will be used to screen for potential EBIs.

3. In the "Population" drop-down menu, select the "GEMA" population for viewing in the image gallery.

4. In the "View" drop-down menu, select "Ch01/Ch05/Ch07/Ch11" to only view this composite image for all events in the image gallery.

5. Click "Tag Objects" and select "Create new" to create a new bin.

6. Manually screen through each photomicrograph in the image gallery and double click to tag "Potential EBIs", which consist of an F4/80$^+$ central macrophage surrounded by 3 or more Ter119$^+$ erythroblasts as shown in Fig. 2 (*see* **Notes 33** and **34**).

7. Click "Save" and name the bin as "Potential EBIs."

3.6 IFC Manual Screening of Potential EBIs for True EBIs

1. Click on the "Image Properties" button and the "Composite" tab. Click "New" to create a new composite image. Click "Add Image" and select "Ch01" (Brightfield 1) at 10%, "Ch03" (CD71-PE) at 100%, "Ch07" (Hoechst33342) at 100%, and "Ch11" (F4/80-APC) at 100% to create the new composite image "Ch01/Ch03/Ch07/Ch11." This composite image will be used together with "Ch01/Ch05/Ch07/Ch11" to identify Ter119$^{low-neg}$ CD71$^+$ proerythroblasts.

2. Click on the "Image Properties" button and the "Composite" tab. Click "New" to create a new composite image. Click "Add Image" and select "Ch01" (Brightfield 1) at 10%, "Ch04" (CD11b-PE-CF594) at 100%, "Ch06" (VCAM1-PE-Cy7) at 100%, "Ch07" (Hoechst33342) at 100%, "Ch11" (F4/80-APC) at 100%, and "Ch12" (Ly6G-APC-Cy7) at 100% to create the new composite image "Ch01/Ch04/Ch06/Ch07/Ch11/Ch12." This composite image will be used to identify non-macrophage myeloid cells.

3. Click on the "Image Properties" button and the "Views" tab. Click "New" to create a new viewing setting for objects in the image gallery. Click "Add Column", check the "Image" box, and include all active channels (as listed in Subheading 3.3). Check the "Composite" box and include all three composite images created above. Rename this viewing option as "All channels + 3 composite."

Fig. 1 (continued) were enriched by gating on Hoechst33342$^+$ F4/80$^+$ Ter119$^+$ CD169$^+$ CD71$^+$ VCAM1$^+$ aggregates based on fluorescence intensity feature and designated as gated erythroid-macrophage aggregates (GEMAs). Intensity positive gates were set far above unstained or FMO controls [9]. Note a tighter intensity gate for F4/80^{hi} Ter119^{hi} aggregates in (**c**) will result in higher enrichment of EBIs, which will reduce subsequent manual screening duration

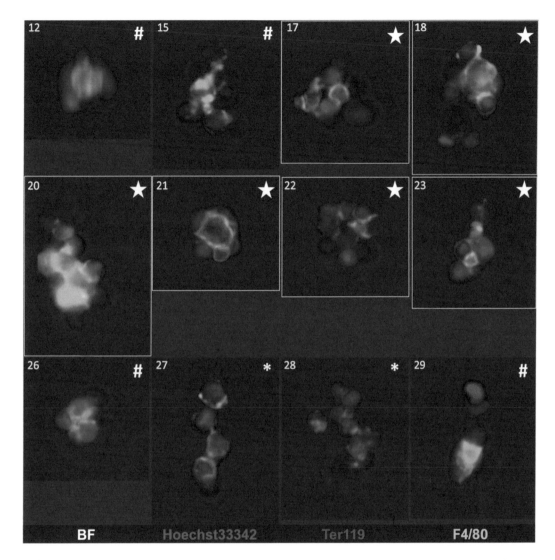

Fig. 2 Examples of potential EBI selection from GEMAs by IFC manual screen. Depicted are composite images "Ch01/Ch05/Ch07/Ch11" of GEMAs in the image gallery at **step 6** in Subheading 3.5. The object number is shown in yellow on the top left of each composite image, with the color legend shown at the bottom. All photomicrographs in this figure were obtained with identical display settings. From these GEMAs, potential EBIs are selected, defined as possessing an F4/80$^+$ central macrophage surrounded by 3 or more Ter119$^+$ erythroblasts. Star (☆): objects 17, 20–23 are potential EBIs as they fit the criteria above with flat, elongated, or contorted F4/80$^+$ cell suspected of being a macrophage. Object 21 is almost certainly a true EBI. Hash (#): objects 12, 15, 26, and 29 appear to contain an F4/80$^+$ cell, but do not contain 3 or more Ter119$^+$ erythroblasts. Asterisk (*): objects 27 and 28 contain 3 or more Ter119$^+$ erythroblasts but possess only F4/80$^+$ fragments

4. In the "Population" drop-down menu, select the "Potential EBI" population for viewing in the image gallery.

5. In the "View" drop-down menu, select "All channels + 3 composite" to change to this viewing option for all events in the image gallery.

6. Click "Tag Objects" and select "Create new" to create a new bin.

7. Manually screen through each event in the image gallery and double click to tag "True EBIs", which consist of an F4/80$^+$ CD169$^+$ VCAM1$^+$ CD11b$^-$ Ly6G$^-$ central macrophage surrounded by 5 or more Ter119$^+$ CD71$^+$ or Ter119$^+$ CD71$^-$ erythroblasts (including Ter119$^{\text{low-neg}}$ CD71$^+$ proerythroblasts) as shown in Figs. 3 and 4 (*see* **Notes 34–38**).

8. Click "Save" and name the bin "True EBIs."

3.7 IFC Quantification of True EBIs

1. Click "New Statistics Table" to view the counts of every population.

2. Under the "File" tab, click on "Save as Template (.ast)" to save a template of the plots with gates and image display settings above.

3. Apply the template (.ast file) and compensation matrix (.ctm file) to the other samples (.rif files).

4. Check for similar proportions and fluorescence intensity profile of events in every gate of the other samples (*see* **Note 39**).

5. Repeat **steps 3–7** in Subheading 3.5 and **4–8** in Subheading 3.6 to identify and count true EBIs in the other samples.

6. Export the counts from every sample to Microsoft Excel.

7. Calculate the number of true EBIs per 10^7 femoral BM cells ($N_{10\text{mil}}$) or the number of true EBIs per femur (N_{femur}) as follows (*see* **Note 40**):

$$N_{10\text{mil}} = \frac{N_{\text{raw}}}{P_{\text{analyzed}}} \text{ or } N_{\text{femur}} = \frac{N_{\text{raw}}}{P_{\text{femur}} \times P_{\text{analyzed}}},$$

where

N_{raw} = raw number of true EBIs counted on IDEAS software (from **step 6** in Subheading 3.7).

P_{analyzed} = proportion of BM sample acquired by volume (from **step 8** in Subheading 3.3).

P_{femur} = proportion of femur (by volume) stained with EBI panel (from **step 8** in Subheading 3.1).

8. In addition to the quantification of EBIs in mouse bone marrow, this method can be adapted to study (a) the frequency of EBIs in other organs such as spleen or fetal liver or organisms; (b) the localization of novel markers present on central macrophages and/or erythroblasts (by substituting antibodies); (c) the shape (using Circularity and other texture features) and size (using Area feature) of the central macrophage, the

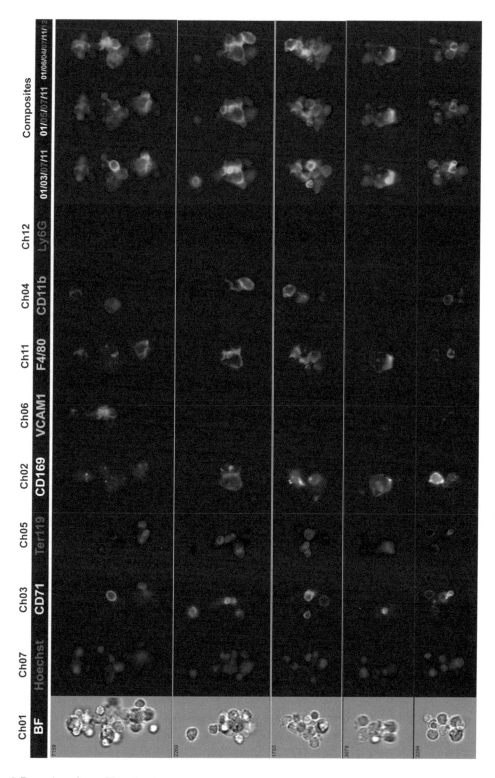

Fig. 3 Examples of true EBI selection from potential EBIs by IFC manual screen. Depicted are potential EBIs viewed with "All channels + 3 composite" option in the image gallery at **step 7** in Subheading 3.6. All

erythroblasts, or the EBIs; (d) the number of erythroblasts per EBI; (e) the presence and interaction of myeloid cells with EBIs; and much more.

4   Notes

1. EBI frequency in the bone marrow has been reported to increase with age in mice [8] therefore it is important to have age-matched mice when comparing EBIs frequencies between mice or between treatment groups.

2. We utilize IMDM because it contains 2 mM Ca^{2+} and 1 mM Mg^{2+} ions which are required to maintain adhesive interactions between macrophage and erythroblasts and maintain integrity of EBIs [15] until the PFA fixation step.

3. We routinely use a 70 μm cell strainer to remove large unseparated cell clumps from triturated mouse bone marrow cell suspensions. A cell strainer with a different pore size may need to be tested for other species such as humans, which have larger cells.

4. 2.4G2 rat anti-mouse CD16/32 neutralizing antibody can be purchased purified from BD Biosciences as Mouse BD Fc Block™. Alternatively, 10% normal rat serum can be used as a blocking buffer because all fluorescent primary monoclonal antibodies are rat immunoglobulins G. Species selection for blocking buffer serum will depend on the species of antibodies used.

5. This antibody panel can only be run on Amnis ImageStream^X Mk II correctly configured with 375/405 nm, 488 nm, and 642 nm lasers and channels configured as shown in **step 1** in Subheading 3.3. Thus our antibody panel has been designed for (a) detection of antigens with weak signals using FITC conjugated antibodies and antigens with strong signals using PE-CF594 (or PE-Texas Red/PE-Dazzle), PerCP-Cy5.5, PE-Cy7, APC, or APC-Cy7 conjugated antibodies in the appropriate channels, and (b) spectral separation of erythroblast, macrophage, and granulocyte markers to minimize spectral spillover. Alternative combinations of fluorophore-conjugated antibodies or fluorescent molecules

Fig. 3 (continued) photomicrographs in this figure were obtained with identical display settings. The color legend is shown at the top. From these potential EBIs, true EBIs are selected, defined as possessing F4/80$^+$ CD169$^+$ VCAM1$^+$ CD11b$^-$ Ly6G$^-$ central macrophage surrounded by 5 or more Ter119$^+$ CD71$^+$ or Ter119$^+$ CD71$^-$ erythroblasts (images with yellow border). Note that although object 7159 contains 6 Ter119$^+$ erythroblasts, these were evenly split between what appears to be two separate F4/80$^+$ CD169$^+$ VCAM1$^+$ CD11b$^-$ Ly6G$^-$ macrophages

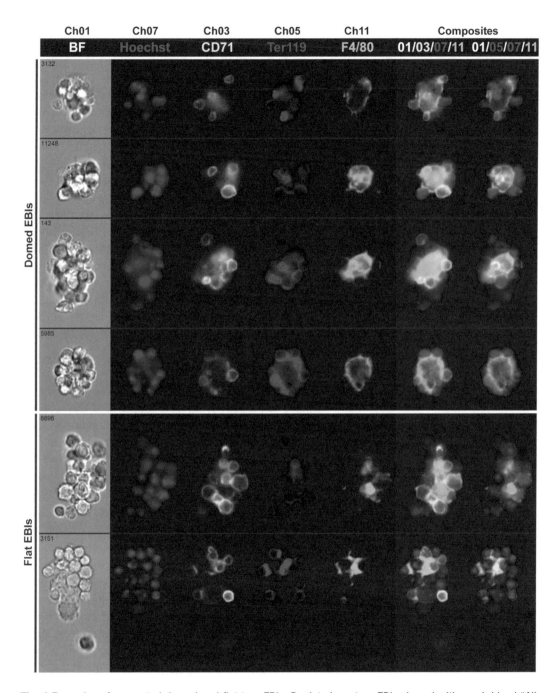

Fig. 4 Examples of suspected domed and flat true EBIs. Depicted are true EBIs viewed with an abridged "All channels + 3 composite" option in the image gallery. All photomicrographs in this figure were obtained with identical display settings. Domed EBIs tend to have central macrophages with clear, round F4/80 membrane staining with most of its associated erythroblasts positioned around the macrophage. Flat EBIs tend to have central macrophages with diffuse, pointy F4/80 membrane staining with some of its associated erythroblasts positioned on top of or below the macrophage

may be used, dependent on the fluorophores supported by the configuration of the imaging flow cytometer used. Since the identification of EBIs requires at least one reliable central macrophage marker (e.g., F4/80) and erythroblast marker (e.g., Ter119), antibodies targeting other antigens may be substituted to study other antigens of interest. Standard flow cytometry antibody panel design guidelines should always be adhered to.

6. For our analyses we utilized a two-camera Amnis ImageStream^X Mk II imaging flow cytometer configured with 405, 488, and 642 nm lasers, INSPIRE acquisition software, and IDEAS analysis software (version 6.0). Some modifications (for example, channel selection) may be required depending on the model of an imaging flow cytometer. For the Amnis FlowCyte instrument the narrower image width compared to the ImageStream^X may limit viewing of EBIs embedded within larger cell aggregates. For the 1-camera Amnis ImageStream^X, the antibody panel will need to be adapted to the limited channels available depending on machine configuration.

7. We use predominantly cervical dislocation as it is quick, but other methods are also compatible with femoral bone marrow collection and may be used.

8. We recommend inserting the needle deeper into the marrow cavity while flushing gently to expel most of the marrow. This method limits the passage of cells through the smaller 23G needle, preventing excess sheer stress and EBI fragmentation.

9. Be gentle when passing the cell suspension through the pipette tip. We use 1 ml pipettes to triturate to minimize sheer stress because the diameter of the pipette tip is relatively large. Triturate until red marrow particles are mostly invisible. We find this generally requires 5–15 pipetting passes.

10. Fixation with PFA is required to crosslink EBIs and maintain their integrity. It is important however not to over-fix cells as conformational epitopes recognized by some antibodies may be chemically modified by the fixation resulting in poor staining. At the end of this step, the sample pellet should turn white in color.

11. Antibodies from other suppliers may need to be titrated for optimal concentration.

12. We usually run samples with approximately 20–25 µL sample volume to concentrate the sample and minimize the time needed to acquire each sample on the Amnis IFC machine. (*see* **Note 21**). A 20 µL sample typically takes around 40 min to run at low speed.

13. Channels 8 and 10 are left blank as Hoechst33342 has significant spectral emission overlap in those two channels (especially channel 8), and nuclei stained by Hoechst33342 are present in all cells except erythrocytes and reticulocytes. It may therefore be advantageous to substitute Hoechst33342 with 4′,6-diamidino-2-phenylindole (DAPI) for nuclear DNA staining if these two channels are required.

14. CD11b and Ly6G are included to visualize granulocytes present around EBIs and to make sure the central macrophage is devoid of these markers [8, 9], They can be substituted for other markers of interest if needed.

15. A simplified version of this EBI panel can alternatively be collected entirely in camera 1 channels by using 7-aminoactinomycin D instead of Hoechst33342 for nuclear staining in channel 5, Ter119-FITC for erythroblasts in channel 2, and F4/80-PE channel 3.

16. The 785 nm laser for measuring side scatter must be turned off as channels 6 and 12 are in use for PE-Cy7 and APC-Cy7, respectively.

17. These are typical laser intensity settings, and investigators should always adjust laser power on their IFC machines by running the sample with the brightest fluorescence intensity in each channel to prevent signal saturation.

18. We always acquire samples on the 60× objective lens at low speed to obtain photomicrographs with better resolution for easier identification of EBI.

19. This singlet classifier includes red blood cells and white blood cells from mouse and excludes smaller debris and calibration beads, used as the acquisition gate for single-stained color control samples. As the size of single cells for other species may vary, this gate may need adjustment accordingly.

20. This aggregate classifier includes aggregates with >2 cells (for mouse species), used as the acquisition gate for all EBI panel samples. Again, this gate may need adjustment for other species depending on the average leukocyte diameter.

21. This all-cells classifier includes all singlets and aggregates only, used only for troubleshooting purposes.

22. It is important to make sure the sheath fluid is running through the machine before loading a sample to prevent the sample from being returned with an error message claiming the presence of air bubbles or inadequate volume in the sample. In our experience, reloading a sample after it is returned by the machine will result in fewer proportions of cell aggregates and EBIs collected.

23. Be sure to remove all bubbles from the sample (we prefer aspiration using a P10 pipette with slim tips) and ensure there is at least 20 μL sample volume before loading into the ImageStream^X in order to avoid the need to return and reload the sample. If a sample is returned, top up the volume to at least 20 μL and remove all bubbles before reloading.

24. The primary purpose of this file is to check that the percentage of total events that fall in the "cell aggregate" classifier is of the same order of magnitude in all sample files to account for potential inconsistencies in sample preparation which will affect the quantification of EBIs. Typically, this value is 0.5–2% in our EBI panel-stained samples from mouse bone marrow.

25. Due to the long acquisition time and processing limitations presented by large data files, we typically only save 10,000 aggregate events per sample.

26. The recording of sample volume before and after acquisition is required for calculating the proportion of sample acquired, and for accurate quantification of EBIs.

27. Alternatively, antibody capture beads may be used for single-stained color controls for compensation except for Hoechst33342 control, which must be done using fixed cells. We prefer to use cells for color controls as the autofluorescence of cell aggregates vary considerably from that of beads.

28. Although automatic spectral compensation on IDEAS tends to work well, it is advisable to always manually check that compensation has been done properly in each single-stained color control file, especially when there is significant spectral overlap between multiple fluorophores in the panel.

29. A focus gate (i.e., gradient RMS feature) is not used here because at least a portion of the three-dimensional cell aggregate will be out of focus, and thus aggregates will have a low gradient RMS score on average compared to singlets.

30. We found that EBIs are exclusively F4/80^{hi} Ter119^{hi} [9], so always apply a high-intensity threshold for these two markers.

31. The pixel intensity negative threshold for each channel can be set on the PFA-fixed unstained sample or fluorescence minus one control. We usually set the positive threshold for pixel intensity on the max pixel intensity of a random event with above-average fluorescence intensity in that channel. Make sure the pixel intensity thresholds are set for optimized viewing of the average aggregate to minimize missing EBIs with lower staining intensities.

32. In our experience, the cross-camera alignment between camera 1 channels (Ch 1–6) and camera 2 channels (Ch 7–12) is not perfect for all events and can deviate significantly in both the x and y planes. When selecting representative composite images for publication, you may want to filter out events with poor alignment between camera 1 and 2 brightfield channels (1 and 9). This can be done by gating on events with −0.2 to 0.2 μm for both Shift X and Shift Y features comparing camera 1 and 2 brightfield images. The Shift X or Shift Y feature reports the shift required in the x and y axis, respectively, to obtain the highest cross-correlation between images from 2 channels (i.e., perfect alignment = 0 for both Shift X and Shift Y). We typically get ~80% of events in the Shift X, and Y gate mentioned above.

33. As the cell aggregate does not adhere to a surface but flows downwards when the photomicrograph is taken, the central macrophage of an EBI can appear in all kinds of shapes and sizes depending on its orientation and distance relative to the camera (see **Note 34** and Fig. 4). In this initial screen, we recommend tagging GEMAs with 3 or more Ter119$^+$ erythroblasts even if they do not have clear F4/80 staining outline of the central macrophage, as it may be a flat EBI.

34. Five or more erythroblasts were set as a criterion for true EBIs as the majority of EBIs from human and mouse have been reported to contain 5 or more erythroblasts [13, 16].

35. We recommend screening through a few thousand GEMAs and potential EBIs in search of true EBIs per sample, which can take up to 1 hour for the trained eye. To reduce screening time for sample GEMAs, the user may screen through half of all GEMAs or potential EBIs and calculate the total number of true EBIs in the .daf file.

36. Note that central macrophages can appear flat or domed [12] as shown in Fig. 4. At least in steady state, the central macrophage is CD11b$^-$ Ly6G$^-$ [8, 9] but may change in stress states.

37. Look out, especially for proerythroblasts, which are devoid of myeloid markers, Hoechst33342$^+$ CD71$^+$ Ter119$^{low-neg}$, and easy to miss when counting the number of erythroblasts.

38. Granulocytes and other CD11b$^+$ myeloid cells are frequently found in close proximity to the central macrophage of EBIs in the bone marrow [9]. It is unclear whether they play a direct role in regulating erythropoiesis.

39. As the intensity of fluorophores may vary slightly across experiments, the "Image Display Mapping" thresholds may need adjusting but should be kept constant for all samples within the same experiment.

40. In our hands, we typically count ~1000 true EBIs per 10^7 femoral BM cells from 8 to 12-week-old male C57BL/6 mice in steady state [9]. This number is likely to be an underestimate as the flushing, trituration, and washing steps are likely to break up some of the EBIs. Deviation from this number of EBIs can be attributed to many factors, including organism, strain, age, sex, housing conditions, in vivo treatments, as well as variation in method and materials used for flushing bone marrow, trituration of marrow, fixation of a sample, staining of a sample, configuration of IFC machine, and analyses of EBI using software.

Acknowledgments

The authors thank the Translational Research Institute, The University of Queensland, and Mater Foundation for providing an excellent research environment and Flow Core Facility. The development of this methodology was supported by funding from the National Health and Medical Research Council Research Fellowship 1136130 (J.P.L.), and Mater Foundation (J.P.L.).

References

1. Erbani J, Tay J, Barbier V, Levesque JP, Winkler IG (2020) Acute myeloid leukemia chemoresistance is mediated by E-selectin receptor CD162 in bone marrow niches. Front Cell Dev Biol 8:668. https://doi.org/10.3389/fcell.2020.00668

2. Gülcüler Balta GS, Monzel C, Kleber S, Beaudouin J, Balta E, Kaindl T et al (2019) 3D cellular architecture modulates tyrosine kinase activity, thereby switching CD95-mediated apoptosis to survival. Cell Rep 29: 2295–2306.e2296. https://doi.org/10.1016/j.celrep.2019.10.054

3. Hritzo MK, Courneya J-P, Golding A (2018) Imaging flow cytometry: a method for examining dynamic native FOXO1 localization in human lymphocytes. J Immunol Methods 454:59–70. https://doi.org/10.1016/j.jim.2018.01.001

4. Gautam N, Sankaran S, Yason JA, Tan KSW, Gascoigne NRJ (2018) A high content imaging flow cytometry approach to study mitochondria in T cells: MitoTracker Green FM dye concentration optimization. Methods 134-135:11–19. https://doi.org/10.1016/j.ymeth.2017.11.015

5. Cerveira J, Begum J, Di Marco Barros R, van der Veen AG, Filby A (2015) An imaging flow cytometry-based approach to measuring the spatiotemporal calcium mobilisation in activated T cells. J Immunol Methods 423:120–130. https://doi.org/10.1016/j.jim.2015.04.030

6. Crane GM, Jeffery E, Morrison SJ (2017) Adult haematopoietic stem cell niches. Nat Rev Immunol 17:573. https://doi.org/10.1038/nri.2017.53

7. Levesque J-P, Jacobsen RN, Winkler IG (2017) The role of mesenchymal stem cells in hematopoiesis. In: Atkinson K (ed) The biology and therapeutic application of mesenchymal cells. Wiley, Hoboken, pp 467–480. https://doi.org/10.1002/9781118907474.ch32

8. Seu KG, Papoin J, Fessler R, Hom J, Huang G, Mohandas N et al (2017) Unraveling macrophage heterogeneity in Erythroblastic Islands. Front Immunol 8:1140. https://doi.org/10.3389/fimmu.2017.01140

9. Tay J, Bisht K, McGirr C, Millard SM, Pettit AR, Winkler IG et al (2020) Imaging flow cytometry reveals that granulocyte colony-stimulating factor treatment causes loss of erythroblastic islands in the mouse bone marrow. Exp Hematol 82:33–42. https://doi.org/10.1016/j.exphem.2020.02.003

10. Bessis M (1958) L'îlot érythroblastique. Unité fonctionnelle de la moelle osseuse. Rev Hematol 13:8–11

11. Jacobsen RN, Perkins AC, Levesque JP (2015) Macrophages and regulation of erythropoiesis. Curr Opin Hematol 22:212–219. https://doi.org/10.1097/moh.0000000000000131

12. Yeo JH, Lam YW, Fraser ST (2019) Cellular dynamics of mammalian red blood cell production in the erythroblastic island niche. Biophys Rev. https://doi.org/10.1007/s12551-019-00579-2

13. Li W, Wang Y, Zhao H, Zhang H, Xu Y, Wang S et al (2019) Identification and transcriptome analysis of erythroblastic island macrophages. Blood 134:480–491. https://doi.org/10.1182/blood.2019000430

14. Bisht K, Tay J, Wellburn RN, McGirr C, Fleming W, Nowlan B et al (2020) Bacterial lipopolysaccharides suppress Erythroblastic Islands and erythropoiesis in the bone marrow in an extrinsic and G- CSF-, IL-1-, and TNF-independent manner. Front Immunol 11:2548. https://doi.org/10.3389/fimmu.2020.583550

15. Soni S, Bala S, Gwynn B, Sahr KE, Peters LL, Hanspal M (2006) Absence of erythroblast macrophage protein (Emp) leads to failure of erythroblast nuclear extrusion. J Biol Chem 281:20181–20189. https://doi.org/10.1074/jbc.M603226200

16. Lee SH, Crocker PR, Westaby S, Key N, Mason DY, Gordon S et al (1988) Isolation and immunocytochemical characterization of human bone marrow stromal macrophages in hemopoietic clusters. J Exp Med 168:1193–1198. https://doi.org/10.1084/jem.168.3.1193

Chapter 4

Imaging Flow Cytometry of *Legionella*-Containing Vacuoles in Intact and Homogenized Wild-Type and Mutant *Dictyostelium*

Amanda Welin, Dario Hüsler, and Hubert Hilbi

Abstract

The causative agent of a severe pneumonia termed "Legionnaires' disease", *Legionella pneumophila*, replicates within protozoan and mammalian phagocytes in a specialized intracellular compartment called the *Legionella*-containing vacuole (LCV). This compartment does not fuse with bactericidal lysosomes but communicates extensively with several cellular vesicle trafficking pathways and eventually associates tightly with the endoplasmic reticulum. In order to comprehend in detail the complex process of LCV formation, the identification and kinetic analysis of cellular trafficking pathway markers on the pathogen vacuole are crucial. This chapter describes imaging flow cytometry (IFC)-based methods for the objective, quantitative and high-throughput analysis of different fluorescently tagged proteins or probes on the LCV. To this end, we use the haploid amoeba *Dictyostelium discoideum* as an infection model for *L. pneumophila*, to analyze either fixed intact infected host cells or LCVs from homogenized amoebae. Parental strains and isogenic mutant amoebae are compared in order to determine the contribution of a specific host factor to LCV formation. The amoebae simultaneously produce two different fluorescently tagged probes enabling tandem quantification of two LCV markers in intact amoebae or the identification of LCVs using one probe and quantification of the other probe in host cell homogenates. The IFC approach allows rapid generation of statistically robust data from thousands of pathogen vacuoles and can be applied to other infection models.

Key words *Dictyostelium discoideum*, ImageStream, Imaging flow cytometry, *Legionella pneumophila*, Pathogen vacuole, Phagocytosis, Phagosome, Vesicle trafficking

---

1 Introduction

1.1 Virulence of Legionella pneumophila *and Formation of the Pathogen Vacuole*

More than 65 *Legionella* species are known to date [1]; yet, *Legionella pneumophila* is the clinically most relevant agent of the severe pneumonia Legionnaires' disease. Upon inhalation of contaminated water droplets, the bacteria are phagocytosed by alveolar macrophages and subvert antibacterial cell-autonomous mechanisms to survive, replicate and ultimately destroy the host cell. *L. pneumophila* is a natural parasite of environmental amoebae,

Natasha S. Barteneva and Ivan A. Vorobjev (eds.), *Spectral and Imaging Cytometry: Methods and Protocols*,
Methods in Molecular Biology, vol. 2635, https://doi.org/10.1007/978-1-0716-3020-4_4,
© Springer Science+Business Media, LLC, part of Springer Nature 2023

and the bacteria replicate within both protozoan and mammalian phagocytes using conserved mechanisms [2–4].

The pathogen manipulates the phagosome to avoid fusion with bactericidal lysosomes and instead forms a replicative niche termed the *Legionella*-containing vacuole (LCV) [5, 6]. LCV formation proceeds through an intricate process dependent on the secretion of more than 300 different "effector" proteins by the intracellular multiplication/defective organelle trafficking (Icm/Dot) type IV secretion system (T4SS) [7–10]. The LCV communicates with the endosomal, secretory, and retrograde trafficking pathways and, ultimately, closely interacts with the endoplasmic reticulum (ER) [11–13].

In order to promote our understanding of pathogen vacuole formation in general and virulence of *L. pneumophila* in particular, methods for the quantitative and non-biased analysis of LCV composition and accumulation kinetics are pivotal. Such analysis can be achieved by proteomic profiling of purified, intact LCVs from *D. discoideum*, macrophage-like cell lines, and primary macrophages [14–17] or by imaging flow cytometry (IFC) [18–20] as outlined here.

1.2 LCV Composition in D. discoideum Parental and Mutant Strains

L. pneumophila survives and replicates within *Dictyostelium discoideum*, an often-employed protozoan experimental host, which is an attractive and versatile model because the haploid social amoeba is genetically tractable and easy to handle [4]. In order to answer biological questions about mechanisms underlying LCV formation, the use of *D. discoideum* mutants and comparison to the isogenic parental strain has proven very fruitful.

D. discoideum mutant strains have been widely used to study the role of phosphoinositide (PI) metabolism, vesicle trafficking pathways, and cytoskeleton dynamics for intracellular replication of *L. pneumophila* [4, 21–23]. Accordingly, the deletion of PI 3-kinases (PI3Ks) promotes intracellular bacterial replication and impairs the transition from tight to spacious LCVs [24]. Likewise, the deletion of the PI 5-phosphatase Dd5P4/OCRL (oculocerebrorenal syndrome of Lowe) [20, 25] or the PI 5-kinase PIKfyve [26] promotes intracellular replication of *L. pneumophila*. In contrast, the disruption of PTEN (phosphatase and <u>ten</u>sin homolog), a PI phosphatase antagonizing PI3Ks, reduces the uptake but does not affect the proliferation of *L. pneumophila* [27]. Moreover, *D. discoideum* lacking the Rab7 GTPase-activating protein (GAP) TBC1D5 showed reduced Rab7 on LCVs [18].

We also analyzed in detail the role of the ER-resident large GTPase Sey1/atlastin in LCV formation. Sey1 catalyzes homotypic fusion of ER tubules and is, therefore, an important regulator of ER morphology and dynamics. In the context of *L. pneumophila* infection, Sey1 promotes ER accumulation at the LCV, expansion of the vacuole positive for the PI lipid phosphatidylinositol 4-phosphate (PtdIns(4)P), as well as intracellular replication of *L. pneumophila* [19, 28].

D. discoideum amoebae are readily transformed with plasmids for ectopic production of fluorescently tagged eukaryotic proteins, *L. pneumophila* effectors, or specific probes [4, 26, 29–32]. The production of two separate fluorescent probes by previously described protocols [33–35] enables tandem IFC analysis of two LCV-localized molecules at once in intact cells [19], or the use of one probe as an LCV marker and quantification of the other [18, 20].

The relative amount of an LCV marker present on the pathogen vacuole as compared to the remainder of the cell varies greatly depending on the protein. Canonical Icm/Dot-dependent LCV markers such as PtdIns(4)P, the ER-resident protein calnexin, or the small GTPase Rab1 are present in high enough relative abundance on the LCV to enable analysis in intact host cells [19, 20, 34]. On the other hand, small relative amounts of LCV marker proteins present on the LCV, such as the retromer coat complex components, necessitate homogenization of the infected cell before analysis, in order to increase the signal-to-noise ratio and to allow detection by either confocal microscopy or IFC [18, 20, 25, 29]. Given these constraints, we here present methods for the quantitative analysis of LCVs in both intact and homogenized amoebae. For the homogenized amoebae, one of the fluorescent probes is used for the identification of the LCVs, while the other – the protein of interest – is quantified on the pathogen vacuole.

1.3 Imaging Flow Cytometry Analysis of Pathogen Vacuole Formation

The quantification of LCV markers is routinely performed by manual scoring of confocal microscopy images depicting pathogen vacuoles of stained or probe-producing infected host cells [24, 25, 29, 30]. This approach is laborious and time-consuming and may introduce a selection bias. In contrast, IFC allows the unbiased quantification of a large range of parameters, exploiting both the spatial resolution of fluorescence microscopy and the high throughput of flow cytometry [18–20].

The ImageStreamX MkII IFC system allows simultaneous analysis of brightfield images and up to 10 fluorescence channels. When using a 60× objective, the resolution of the system is 0.33 μm with a 2.5 μm depth of field. In theory, thousands of cells can be analyzed per second, although in practice and with an optimal sample concentration, 10'000 cells are loaded and acquired in approximately one minute. After color compensation, an analysis strategy is followed, where masks are created that define the area of interest in the cell. Subsequently, pre-existing or user-made so-called "features" are applied, yielding a statistically robust quantification of a large range of parameters that can be applied to all samples using a template. This workflow allows rapid, highly resolved, and unbiased quantification of intracellular events, with a range of applications including pathogen vacuole formation.

In this chapter, we describe how IFC can be used to quantify the localization of fluorescently tagged proteins on LCVs in intact or homogenized *L. pneumophila*-infected *D. discoideum* amoebae. One of the quantified LCV markers is the PI lipid PtdIns(4)P, a major regulator of secretory vesicle trafficking through the Golgi apparatus. PtdIns(4)P localizes to the plasma membrane and accumulates in an Icm/Dot-dependent manner on the LCV, where it replaces the endocytic PI lipid PtdIns(3)P [23, 24, 33, 34]. The accumulation can be assessed with the PtdIns(4)P probe P4C$_{SidC}$, comprising the unique 20 kDa PI-binding domain of the *L. pneumophila* Icm/Dot substrate SidC [36]. The P4C$_{SidC}$ probe can be quantified on LCVs in tandem with a second marker in intact cells by IFC [28] and can also be used to identify LCVs in host cell homogenates.

Another robust LCV marker is calnexin, an ER-resident protein that accumulates in an Icm/Dot-dependent manner around the LCV as the infection progresses [19, 36–38]. Calnexin accumulation is considered a hallmark of productive LCV formation, and the ER protein does not decorate vacuoles harboring Icm/Dot-deficient *L. pneumophila* mutant strains. In addition to studying pathogen vacuole formation upon infection with *L. pneumophila*, the presented method can be adapted for other pathogens of interest, including *Mycobacterium* spp. [39].

2 Materials

2.1 Legionella Pneumophila *Strains*

1. *L. pneumophila* virulent Philadelphia-1 strain JR32 [40] harboring the mPlum-producing plasmid pAW14 [19].

2. *L. pneumophila* isogenic avirulent mutant strain GS3011 (Δ*icmT*, JR32 *icmT*3011::Kan), lacking a functional Icm/Dot T4SS [41] and harboring the mPlum-producing plasmid pAW14 [19].

2.2 *Culture Medium* for L. Pneumophila *Strains*

1. ACES yeast extract (AYE) medium [42]: 10 g/L N-(2-acetamido)-2-aminoethanesulfonic acid, 10 g/L Bacto™ yeast extract (*see* **Note 1**), 0.6 mM Fe(NO$_3$)$_3$, 3.3 mM L-cysteine. Add 10 g of ACES and 10 g of yeast extract in 950 mL of H$_2$O. Add 0.25 g/10 mL Fe(NO$_3$)$_3$ solution and filter-sterilized 0.4 g/10 mL L-cysteine (*see* **Note 2**). Adjust with 10 M KOH the pH to 6.9. Pass the medium several times through a glass fiber filter, and then through a 0.2 μm filter cartouche. Store the medium in the dark at 4 °C (*see* **Note 3**). To select for plasmid pAW14, add 5 μg/mL chloramphenicol (Cam, stock: 30 mg/mL ethanol).

2. Prepare CYE (charcoal yeast extract) agar plates [43] as follows: 10 g/L ACES, 10 g/L Bacto™ yeast extract (*see* **Note 1**), 15 g/L agar, 2 g/L activated charcoal powder (puriss. p.a.), 3.3 mM L-cysteine, 0.6 mM Fe(NO$_3$)$_3$. Dissolve 10 g of yeast extract and 10 g of ACES in 950 mL of H$_2$O and adjust the pH to 6.9. Transfer the solution to a 1-L Schott bottle containing 15 g of agar and 2 g of activated charcoal powder, and a stir bar. Autoclave and let the agar solution cool down to 50 °C. Add 0.4 g/10 mL L-cysteine and 0.25 g/10 mL Fe(NO$_3$)$_3$ filter-sterilized solutions (*see* **Note 2**). To prepare plates to select for plasmid pAW14 add 5 µg/mL Cam. Mix the solution with a stir bar on a magnetic stirrer and pour plates. Dry plates to remove condensation water and store at 4 °C for up to 6 months.

2.3 *Instrumentation*

1. Spectrophotometer: Lambda XLS (PerkinElmer).

2. Sterile 13 mL test tubes with ventilated cap.

3. Inverted light microscope.

4. Rotation wheel.

2.4 **Dictyostelium Discoideum** *Strains*

1. The axenic *D. discoideum* strain Ax3 [44] and its isogenic mutant Δ*sey1* [28], both transformed with pWS32 for the production of the PtdIns(4)P probe P4C$_{SidC}$-mCherry [19] and in tandem with pAW16 for production of calnexin-GFP (green fluorescent protein) [18], as well as strain Ax3 transformed with each construct individually, are used (*see* **Note 4**).

2.5 **Dictyostelium Discoideum** *Medium*

1. HL5 medium, modified [45] (*see* **Note 5**): 5 g/L BBL™ yeast extract (BD Difco); *see* **Note 1**), 5 g/L Bacto™ Proteose Peptone (BD Difco; *see* **Note 1**), 5 g/L BBL™ Thiotone™ Peptone (BD Difco; *see* **Note 1**), 11 g/L D(+)glucose monohydrate (*see* **Note 6**), 2.5 mM Na$_2$HPO$_4$, 2.5 mM KH$_2$PO$_4$. Adjust the pH to 6.5. Autoclave and store the medium at 4 °C. The modified HL5 medium supports axenic growth of *D. discoideum*. If necessary, add penicillin/streptomycin or Fungizone to maintain sterility.

2.6 *Additional Instrumentation and Stock Solutions*

1. Antibiotics for cell culture: 20 µg/mL G418 (stock: 20 mg/mL dH$_2$O), 50 µg/mL hygromycin (stock: 50 mg/mL dH$_2$O). Filter-sterilize and store the stock solutions at 4 °C (hygromycin) or −20 °C (G418).

2. Tissue culture flasks (75 cm^2).

3. Incubators set to 23 °C and 25 °C.

4. Hemocytometer.

2.7 Experimental Infections. Intact Host Cells

1. 12-well tissue culture plates.

2. Cell culture centrifuge with swing-out rotors for plates.

3. Microcentrifuge.

4. DPBS (Dulbecco's phosphate-buffered saline).

5. Dilute 16% PFA (paraformaldehyde) in DPBS to prepare 2% solution, and store at -20 °C.

2.8 Experimental Infections. Homogenized Host Cells

1. 75 cm^2 tissue culture flasks.

2. Sorensen phosphate buffer [46] with CaCl$_2$ (SorC): 2 mM Na$_2$HPO$_4$, 50 µM CaCl$_2$, 15 mM KH$_2$PO$_4$. Adjust the pH to 6.0, autoclave, and store at room temperature. Add 1 mL of a sterile 50 mM CaCl$_2$ stock solution.

3. HEPES-sucrose homogenization buffer: 20 mM N-2-hydroxyethylpiperazine-N'-2-ethanesulfonic acid (HEPES), 0.5 mM ethyleneglycoltetraacetic acid (EGTA), 250 mM sucrose. Adjust the pH to 7.2, sterilize through a 0.2-µm filter, and store at 4 °C for up to 6 months. Prior to use, supplement with cOmplete™ protease inhibitor cocktail tablets (Sigma, cat. no. 11836170001).

4. Plastic cell scraper.

5. 13 mL test tubes with caps.

6. Stainless steel ball homogenizer, 8-µm clearance, 0.5-mL chamber (Isobiotec).

7. Plastic Luer-Lok™ syringes.

2.9 Imaging Flow Cytometry

1. 1.5 mL Eppendorf microcentrifuge tubes.

2. ImageStreamX MarkII system with Inspire software for data acquisition (Amnis-Luminex).

3. IDEAS 6.2 software for data analysis (Amnis-Luminex).

---

3 Methods

3.1 Culture of L. Pneumophila

1. Streak out *L. pneumophila* strains JR32 (wild-type) and Δ*icmT* (Icm/Dot-deficient mutant) constitutively producing mPlum (pAW14) from frozen glycerol stocks (25% glycerol concentration), on a CYE agar plate containing Cam to maintain the plasmid.

2. Incubate for 3 days at 37 °C.

3. Prepare early stationary broth cultures the day before the experiment by inoculating 3 mL AYE broth containing Cam in a test tube (13 mL) with each bacterial strain at optical density 0.1 (600 nm). Incubate on a rotating wheel for 21 hours (approximately 80 rpm, ca. 1 × g) at 37 °C (*see* **Note 7**).

3.2 Culture of D. Discoideum

1. Start *D. discoideum* Ax3 (parental strain) and Δ*sey1* cultures containing the necessary constructs from frozen stocks by rapidly thawing the aliquots and inoculating 10 mL HL5 medium at room temperature (RT) in 75 cm^2 flasks.

2. Incubate the flasks at 23 °C for 1–2 h to allow the cells to attach to the surface.

3. Exchange the HL5 medium for removing the freezing medium and adding the appropriate antibiotics (*see* **Note 8**).

4. Culture cells at 23 °C for 2–4 days until 70–80% confluency is reached.

5. Sub-culture cells into 75 cm^2 flasks (*see* **Note 9**) and dilute approximately 1:100 in fresh HL5 medium (*see* **Note 10**).

6. The day prior to infection, seed *D. discoideum* Ax3 and Δ*sey1* producing simultaneously P4C$_{SidC}$-mCherry and calnexin-GFP in a 12-well plate (5×10^5 per well) for analysis of host cells or in 75 cm^2 flasks (1×10^7 per flask) for analysis of homogenized amoebae (*see* **Note 11**). Also, sub-culture *D. discoideum* Ax3 producing only P4C$_{SidC}$-mCherry in one well, and *D. discoideum* Ax3 producing only calnexin-GFP in another well. These amoebae strains will be used as color compensation controls.

3.3 Experimental Infections for the Analysis of Intact Host Cells

1. Remove *L. pneumophila* strains JR32 and Δ*icmT* producing mPlum from the rotating wheel.

2. Determine the concentration of *L. pneumophila* (*see* **Note 7**).

3. Dilute the suspension in HL5 medium (w/o antibiotics) to 5×10^6/mL, for a multiplicity of infection (MOI) of five (*see* **Note 12**).

4. Remove the medium from the 12-well plate containing *D. discoideum* Ax3 and Δ*sey1* strains producing P4C$_{SidC}$-mCherry and calnexin-GFP (*see* **Note 13**).

5. Add 1 mL of the bacterial suspension to the wells containing *D. discoideum* Ax3 and Δ*sey1* producing P4C$_{SidC}$-mCherry and calnexin-GFP. For mPlum color compensation control take 1 mL of the suspension in a 1.5 mL microcentrifuge tube.

6. Centrifuge the plate for 10 min at $450 \times g$ (RT) to synchronize the uptake of the bacteria by *D. discoideum*.

7. Incubate the plate at 25 °C (*see* **Notes 14** and **15**).

8. At each time point, remove the plate from the incubator and resuspend the cells. Transfer the samples to pre-labeled 1.5 mL tubes. Also, detach the control cells for color compensation.

9. Centrifuge the cells and remove the supernatant (*see* **Note 16**).

10. Gently resuspend the pellet in 200 μl ice-cold 2% PFA and incubate for 90 min on ice.

11. Add 1 mL ice-cold DPBS and repeat step 9.

12. Resuspend the cells in 20 μl ice-cold DPBS (*see* **Note 17**).

3.4 Experimental Infections for the Analysis of Homogenized Host Cells

1. Remove tubes with *L. pneumophila* strain JR32 producing mPlum from the rotating wheel and determine the concentration of bacteria (*see* **Note 7**).

2. Dilute the suspension in HL5 medium (w/o antibiotics) to 1×10^8/mL (MOI equals 50) (*see* **Note 18**).

3. Remove the culture medium from the *D. discoideum* Ax3 and Δ*sey1* strains producing P4C$_{SidC}$-mCherry and calnexin-GFP grown in the 75 cm^2 flasks.

4. Add 10 mL of the diluted bacterial suspension to the flasks containing *D. discoideum* Ax3 or Δ*sey1* strains, producing P4C$_{SidC}$-mCherry and calnexin-GFP. For mPlum color compensation control take 1 mL of the bacterial suspension in a 1.5 mL microcentrifuge tube (*see* **Note 19**).

5. Centrifuge the flasks for 10 min at $450 \times g$ (RT) to synchronize uptake of the bacteria by amoebae.

6. Incubate the flasks at 25 °C for the different time points (*see* **Note 14**).

7. Take the flasks out from the incubator and place them on ice. Wash the infected *D. discoideum* cells by gentle agitation changing the medium (10 mL of cold SorC) three times.

8. Add 3 mL of ice-cold HS homogenization buffer to each flask and collect the cells into a 15 mL test tube by using a plastic cell scraper.

9. Wash the ball homogenizer once with ethanol, then thoroughly with distilled water, and finally, flush the chilled homogenizer with cold HS buffer to remove air bubbles.

10. Transfer the 3 mL of *L. pneumophila*-infected *D. discoideum* cells into a disposable Luer Lock syringe. Mount the syringe on the homogenizer and press the suspension through the homogenizer into a second syringe. Passage the suspension back and forth seven to nine times.

11. Keep the homogenate on ice. Disassemble and thoroughly wash the ball homogenizer again before proceeding with each sample.

12. Centrifuge the homogenates at $2700 \times g$ for 15 min at 4 °C. Discard the supernatant completely and re-suspend the pellet in 200 μl ice-cold 2% PFA by gently pipetting.

13. Incubate on ice for 1 h.

14. Add 1 mL ice-cold DPBS, centrifuge, and pellet the homogenates as in step 12.

15. Remove the supernatant and resuspend the homogenates in 20 μl ice-cold DPBS (*see* **Notes 17** and **20**).

3.5 Acquisition of Imaging Flow Cytometry Data on Intact Host Cells

1. Turn on the ImageStreamX MarkII system and run the ASSIST calibrations and tests.

2. Choose the objective 60× and select low speed (*see* **Note 21**).

3. Set brightfield to Ch01 (*see* **Note 22**).

4. Turn on the 488 nm and 561 nm lasers for excitation of GFP, mPlum, and mCherry, and the side scatter (SSC; 785 nm) laser for darkfield observation (*see* **Note 23**).

5. Create a histogram for [Gradient RMS_M01_Ch01]. Set a gate (R1) from 50 and up (*see* **Note 24**).

6. Create a dot plot with [Area_M01] vs. [Aspect Ratio_M01] including only gated events. Set gate (R2) around Area 50–200 and Aspect Ratio 0.5–1 (*see* **Note 25**).

7. Create a dot plot with [Intensity_MC_Ch02] on the x-axis vs. [Raw Max Pixel_MC_Ch02] on the y-axis, including only the events in R2. Repeat for each fluorescence channel (Ch04, Ch05, and Ch06) (*see* **Note 26**).

8. Load the first sample (*see* **Note 27**).

9. Adjust R1 and R2 gates to include only single-focused events (*see* **Note 28**).

10. Use the Intensity versus Raw Max Pixel plots to adjust the power of the three lasers (*see* **Note 29**).

11. Acquire 10,000 cells in the R2 gate for each sample.

12. Turn off the brightfield and the side scatter, and set the system to acquire the "All" population. Collect 5000 events for each color compensation control.

13. Sterilize and shut down the system.

3.6 Acquisition of Imaging Flow Cytometry Data from Homogenized Host Cells

1. Repeat steps 1–4 from Subheading 3.5.

2. Create a histogram with [Gradient RMS_M05_Ch05] on the x-axis. Set a line region (R1) from around 40 and up (*see* **Notes 28** and **30**).

3. Create a dot plot with [Intensity_MC_M04] on the x-axis vs. [Intensity_MC_M05] on the y-axis, including only the events in R1. Set a gate (R2) around [Intensity_MC_M04] 5e3-1e6 and [Intensity_MC_M05] 3e3-1e6 (*see* **Note 31**).

4. Create a dot plot with [Intensity_MC_Ch02] vs. [Raw Max Pixel_MC_Ch02] on the y-axis, including all events. Repeat for each fluorescence channel (Ch04, Ch05, and Ch06) (*see* **Note 26**).

5. Go to the Advanced Setup menu and then the Acquisition tab. Set Squelch to around 60 (*see* **Note 32**).

6. Load the first sample (*see* **Note 27**).

7. Use the Intensity versus Raw Max Pixel plots to adjust the power of the three lasers (*see* **Note 29**).

8. Return the first sample (*see* **Note 33**).

9. Open the compensation wizard and follow the steps to create and apply a compensation matrix (*see* **Note 34**).

10. Re-load the first sample.

11. Adjust R1 and R2 as necessary. Try to include only cells with bacteria in focus (R1) and clearly positive for both mCherry and mPlum (R2).

12. Acquire 20,000 events in the R2 gate for each sample.

13. Sterilize and shut down the system.

3.7 Analysis of Imaging Flow Cytometry Data of Intact Host Cells

1. Click on "Start Analysis" in IDEAS software to open the first file and apply a compensation matrix created by using the three collected color compensation controls. All steps included in the analysis of fluorescent marker accumulation on LCVs in intact cells are depicted in Fig. 1. Each step includes only the events gated in the previous step (*see* **Note 35**).

2. Follow the "Internalization Wizard" to identify the in-focus, individual amoebae double positive for mCherry and GFP that have internalized *L. pneumophila* positive for mPlum (*see* **Note 36**) (Fig. 1a–e).

3. Follow the "Spot Wizard" to create a mask identifying the *L. pneumophila* bacteria and create a histogram of the population of cells with internalized bacteria (identified in step 2). Gate the cells having internalized only a single bacterium (*see* **Note 37**) (Fig. 1f).

4. Follow the "Colocalization Wizard" to create a feature quantifying the colocalization of mPlum and mCherry and use this feature to analyze the population having internalized only one bacterium (identified in step 3) (*see* **Note 38**) (Fig. 1g, left panel).

5. Repeat step 4 for colocalization of GFP and mPlum (Fig. 1g, right panel). The histograms resulting from steps 4–5 show the colocalization of mPlum (*L. pneumophila*) with mCherry (P4C$_{SidC}$) or GFP (calnexin) and demonstrate the relative enrichment of these proteins on the LCVs (Fig. 1g; Fig. 2a).

6. Optionally, create gates on the histograms produced in steps 4–5 for LCVs positive for P4C$_{SidC}$ and calnexin (*see* **Note 39**) (Fig. 1g).

7. Define a statistics report containing the colocalization statistics for GFP and mCherry in the cell population harboring a single LCV (*see* **Note 40**).

8. Follow with Batch analysis using the compensation matrix created in step 1. Open and inspect all the files to adjust gates as necessary.

9. Use GraphPad Prism (Dotmatics, USA) or appropriate statistical software as required (Fig. 2b).

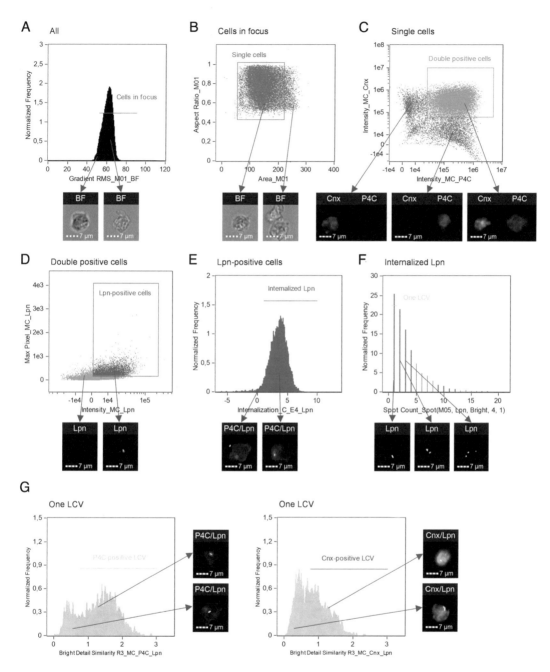

Fig. 1 Gating steps were applied during IFC analysis of LCV markers in intact cells. *D. discoideum* Ax3 dually producing P4C$_{SidC}$-mCherry (pWS32) and calnexin-GFP (pAW16) was infected (MOI 5) for 2 h with mPlum-producing virulent *L. pneumophila* JR32 (pAW14). After fixation, 10′000 cells were acquired with an imaging flow cytometer. The analysis was performed in 7 steps, where the cells gated in one step were carried on to the subsequent step. Examples of included and excluded cells are shown for each step (arrows). (**a**) Cells in focus were gated using the feature [Gradient RMS_M01_BF]. (**b**) Single cells were gated using [Area_M01] vs. [Aspect Ratio_M01]. (**c**) Cells producing both fluorescent proteins were gated using [Intensity_MC_P4C] vs. [Intensity_MC_Cnx]. (**d**) Cells positive for *L. pneumophila* were gated using [Intensity_MC_Lpn] vs. [Max Pixel_MC_Lpn]. (**e**) Cells with internalized *L. pneumophila* rather than adherent

3.8 Imaging Flow Cytometry of Homogenized Host Cells

1. Click on "Start Analysis" in IDEAS software to open the first file. Use the three collected color compensation controls to create and apply a new compensation matrix for the analysis (*see* **Note 41**). All steps included in the analysis of fluorescent marker accumulation on LCVs in host cell homogenates are depicted in Fig. 3. Each step includes only the events gated in the previous step (*see* **Notes 35** and **42**).

2. To gate events where the *L. pneumophila* (mPlum) bacteria are in focus, create a new histogram with [Gradient RMS_M05_Lpn] on the x-axis. Gate the events with a high Gradient RMS (*see* **Note 30**) (Fig. 3a).

3. To exclude larger host cell debris and events where the mPlum signal is not high enough, create a new dot plot of the population identified in step 2 with [Area_MC] on the x-axis and [Max Pixel_MC_Lpn] on the y-axis (Fig. 3b).

4. To exclude images depicting multiple LCVs or long filaments, create a new dot plot of the small bacteria-containing events (identified in step 3) with the Spot Count feature in the appropriate mask, as determined through the Spot Count Wizard, on the x-axis (*see* **Note 43**) and [Area_Object(M05, Lpn, Tight)] on the y-axis (*see* **Note 44**). Gate events containing only one bacterium with a small area (Fig. 3c).

5. To gate the events where the bacteria are surrounded by the marker P4C$_{SidC}$-mCherry (i.e., the LCVs), create a new dot plot of the population identified in step 4 with [Max Pixel_MC_P4C] feature on the x-axis and [Bright Detail Similarity R3_MC_Lpn_P4C] on the y-axis (*see* **Note 45**). Gate events that have a high mCherry signal (i.e., positive for P4C$_{SidC}$) and where this signal colocalizes to a large extent with the signal from the bacterium (Fig. 3d).

6. To quantify the fluorescence intensity of calnexin-GFP on the LCV, representing its abundance, create a histogram of the *L. pneumophila*-containing and P4C$_{SidC}$-mCherry-positive LCV population (identified in step 5) with [Mean Pixel_M02_Cnx] on the x-axis (*see* **Note 46**). (Figs. 3e and 4a).

Fig. 1 (continued) on the surface were gated using [Internalization_C_E4_Lpn]. (**f**) Cells containing exactly one bacterium were selected using [Spot Count_Spot(M05, Lpn, Bright, 4, 1)]. (**g**) The included cells from step F were analyzed for colocalization between a *L. pneumophila* bacterium and P4C$_{SidC}$-mCherry (left panel) using the feature [Bright Detail Similarity R3_MC_P4C_Lpn]. A gate was set at 0.75 to include only cells containing a P4C$_{SidC}$-positive LCV. Similarly, the included cells from step G were analyzed for colocalization between a *L. pneumophila* bacterium and calnexin-GFP (right panel) using [Bright Detail Similarity R3_MC_Cnx_Lpn]. A gate was set at 0.75 to include only cells containing a calnexin-positive LCV

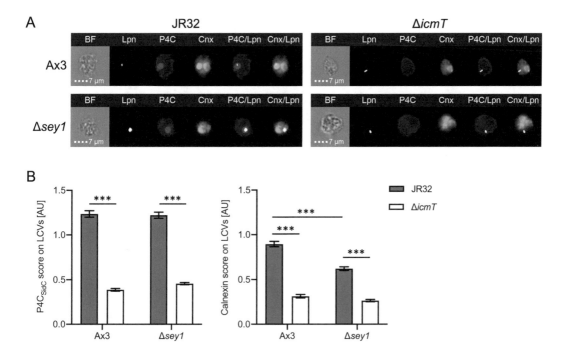

Fig. 2 Simultaneous quantification of PtdIns(4)*P* and calnexin on LCVs in intact parental or Δ*sey1* *D. discoideum* by IFC. *D. discoideum* Ax3 (parental) or Δ*sey1* amoebae were dually transfected to produce P4C$_{SidC}$-mCherry and calnexin-GFP, infected for 2 h with virulent *L. pneumophila* JR32 or Δ*icmT*, fixed and analyzed by IFC (see text for details). (**a**) Representative IFC images showing the colocalization of *L. pneumophila* JR32 or Δ*icmT* (mPlum, white) with P4C$_{SidC}$ (mCherry, red) and calnexin (GFP, green). (**b**) Quantification of P4C$_{SidC}$ (left) and calnexin (right) colocalization with *L. pneumophila*. Representative data for JR32- or Δ*icmT*-infected *D. discoideum* Ax3 or Δ*sey1* at 2 h post-infection (*n* > 500 cells per sample). Data show means and 95% confidence intervals from one experiment (***, *P* < 0.001, regular two-way ANOVA followed by Tukey's posthoc test). Note how P4C$_{SidC}$ has localized to the LCV in an Icm/Dot-dependent but Sey1-independent manner, while calnexin recruitment to the LCV is dependent on both Icm/Dot and Sey1

7. Optionally, set a gate for calnexin-GFP-positive LCVs (Fig. 3e) (*see* **Note 47**).

8. Create a statistics report containing the intensity statistics for GFP in the *L. pneumophila*-containing and P4C$_{SidC}$-mCherry-positive LCV population (*see* **Note 48**).

9. Follow with Batch analysis using the compensation matrix created in step 1. Open and inspect all the files to adjust gates as necessary.

10. Plot the statistics generated in step 10 with statistical software (e.g., GraphPad Prism (Dotmatics, USA) (Fig. 4b).

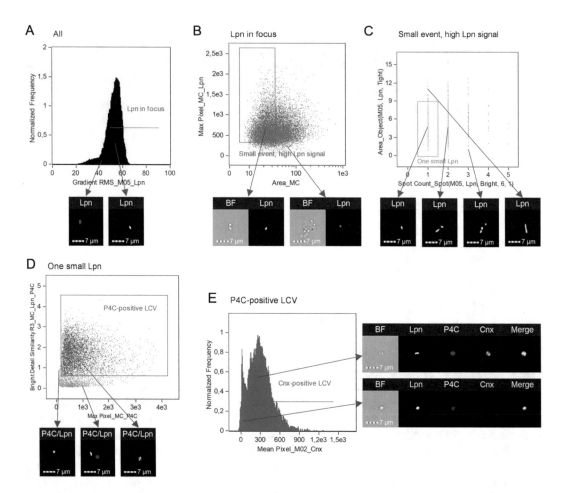

Fig. 3 Gating steps applied during IFC analysis of LCV markers in homogenized host cells. *D. discoideum* Ax3 dually producing P4C$_{SidC}$-mCherry (pWS32) and calnexin-GFP (pAW16) was infected (MOI 50) for 2 h with mPlum-producing virulent *L. pneumophila* JR32 (pAW14). After homogenization and fixation, 20'000 mCherry- and mPlum-positive events were acquired with an imaging flow cytometer. The analysis was performed in 5 steps, where the cells gated in one step were carried on to the subsequent step. Examples of included and excluded events are shown for each step (arrows). (**a**) Events with bacteria in focus were gated using the feature [Gradient RMS_M05_Lpn]. (**b**) Small events with a sufficiently high mPlum signal were gated using [Area_MC] vs. [Max Pixel _MC_Lpn]. (**c**) Events containing exactly one spot corresponding in size to single planktonic *L. pneumophila* were gated using the features [Spot Count_Spot(M05, Lpn, Bright, 6, 1)] vs. [Area_Object(M05, Lpn, Tight)]. (**d**) Events with P4C$_{SidC}$-mCherry in close proximity to the bacterium (i.e., LCVs) were gated using [Max Pixel_MC_P4C] vs. [Bright Detail Similarity R3_MC_Lpn_P4C]. (**e**) The included cells from step D were analyzed for intensity of calnexin-GFP using the feature [Mean Pixel_M02_Cnx]. A gate was set to include only calnexin-positive LCVs

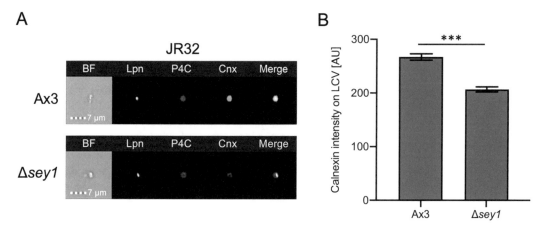

Fig. 4 Quantification of calnexin on LCVs in homogenized parental or Δ*sey1 D. discoideum* by IFC. *D. discoideum* Ax3 (parental) or Δ*sey1* amoebae were dually transfected to produce P4C$_{SidC}$-mCherry and calnexin-GFP, infected for 2 h with mPlum-producing virulent *L. pneumophila* JR32 (pAW14), homogenized, fixed and analyzed by IFC as described in the text for homogenized cells. (**a**) Representative IFC images showing the localization of the LCV marker P4C$_{SidC}$ (mCherry, red) and calnexin (GFP, green) on LCVs containing *L. pneumophila* JR32 (mPlum, white). (**b**) Quantification of calnexin-GFP on LCVs from homogenized, JR32-infected *D. discoideum* Ax3 or Δ*sey1* at 2 h post-infection (*n* > 2000 LCVs per sample). Data show means and 95% confidence intervals from one experiment (***, *P* < 0.001, unpaired Student's t-test). Note how calnexin recruitment to the LCV is dependent on Sey1

4   Notes

1. The quality of yeast extract and peptone affect the physiology of *L. pneumophila* and *D. discoideum*. The yeast extract and peptone should be tested, and the same suppliers/batches should be used for all experiments.

2. Dissolve Fe(NO$_3$)$_3$ and L-cysteine separately in 10 mL of H$_2$O in a 15 mL tube. To prevent precipitation, stir the medium and slowly add the L-cysteine solution first, followed by the iron solution.

3. To remove precipitates, pre-filter the medium 6–8 times through a glass fiber filter paper and store in the dark.

4. Plasmid pWS32 producing P4C$_{SidC}$-mCherry [19]} was cloned in the pDM1044 background [47], has a mCherry tag (at C-terminus), and carries a hygromycin resistance cassette. pAW16 producing calnexin-GFP [18]} was cloned in the pDM323 background [48], has a GFP tag (at C-terminus), and carries a G418 resistance cassette.

5. Alternatively, commercially available HL5 medium (Formedium, UK) may be used.

6. Glucose caramelizes upon autoclaving in combination with the medium. Suspend the D- (+) glucose in 50 mL of pre-warmed H_2O, sterilize (0.2 µm filter), and add to the autoclaved medium.

7. *L. pneumophila* bacteria grown to post-exponential/early stationary growth phase in the AYE medium are morphologically uniform, flagellated, and motile rods (~2 × 0.5 µm). The number of long, filamentous *L. pneumophila* (>20 µm) is much smaller than that in bacterial cultures grown on CYE agar plates. The morphology of the bacteria can be determined by light microscopy (which requires 10 µL of the bacterial culture). The final OD_{600} should be above 5.0 (approximately 2 × 10^9 bacteria/mL); otherwise, the uptake and infection efficiency is severely compromised.

8. To select amoebae containing P4C$_{SidC}$-mCherry, use hygromycin (50 µg/mL). To select amoebae containing pAW16, use G418 (20 µg/mL).

9. *D. discoideum* can be detached by gently tapping the closed flask or by repeated pipetting over the surface.

10. The seeding host cells density should be approximately 2–5 × 10^4/mL. Over-dilution will result in slower replication rates and increased time to reach sub-confluence. Aim to split cultures when the cells are evenly spaced in the exponential growth phase. Never allow the cells to reach confluence, as doing so will drastically reduce transformation efficiency. Exchange the medium every 3 days. Start a new culture from frozen stock, and discard the old culture after a maximum of 3 weeks.

11. The cells have to be seeded at half of the required concentration the day before the experiment. Thus, a seeding density of 5 × 10^5 cells per well will yield 1 × 10^6 cells per well at the time of infection for analysis of intact cells. A seeding density of 1 × 10^7 cells per flask will yield 2 × 10^7 cells at the time of infection for analysis of homogenates.

12. It is important to have a sufficiently high number of infected cells in each sample for optimal speed of IFC analysis. We have determined the optimal MOI for IFC of intact cells to be 5 bacteria per amoeba (i.e., 5 × 10^6 bacteria per well containing 1 × 10^6 amoebae after overnight growth). With this MOI, a high number of cells (usually >1000 of the 10,000 acquired cells after infection with virulent *L. pneumophila*) will remain after all the gating steps, enabling robust statistical analysis.

13. The wells containing *D. discoideum* Ax3 producing only one fluorescent construct (to be used for color compensation) can be left in the original seeding medium throughout the experiment. It is important not to add any bacteria to the color

compensation controls, as these should contain only one fluorescent construct.

14. Herein, incubation of 2 h is used to enable an analysis of mature LCVs. In the case of a time-course experiment assessing the accumulation of LCV markers the starting time corresponds to the time when the plate is put in the 25 °C infection incubator.

15. While incubating, prepare for step 7 by labeling 1.5 mL Eppendorf tubes and thawing and chilling 2% PFA.

16. Intact *D. discoideum* can be pelleted by centrifugation at $500 \times g$ for 5 min at 4 °C or by "quick spin" centrifugation for 10 sec.

17. $1 \times$ DPBS without Ca^{2+} or Mg^{2+} should be used to dilute the samples. Samples can be saved for up to a week at 4 °C in the dark and can be kept at RT during data acquisition.

18. Due to shear stress from homogenization and the fragility of isolated LCVs, homogenized host cells have to be infected with higher MOI (MOI = 50) than intact cells (MOI = 5) to achieve a sufficient LCV yield.

19. The uninfected cells (to be used for color compensation) can be kept in the incubator until the homogenates are fixed in parallel with the experimental samples. The bacterial suspension for color compensation should also be pelleted and processed together with the other color compensation samples.

20. Too concentrated homogenates should be diluted in, e.g., 200 μl $1 \times$ DPBS to allow imaging of single LCVs.

21. Using a lower magnification objective or increasing the speed will affect the quality of images.

22. The channel(s) used for the brightfield image(s) depends on whether a 6- or 12-channel (1- or 2-camera) Imagestream instrument is used. Brightfield can be placed in any free channel. We routinely use channels 1 and 9 for brightfield in a 2-camera 12-channel system.

23. With this setup, GFP, mCherry, and mPlum will be visible in Ch02, Ch04, and Ch05. Darkfield (autofluorescence upon excitation at 785 nm, corresponding to side scatter/granularity) will be in Ch06. Initially, set the lasers to maximum power.

24. Out-of-focus events have low Gradient RMS in the brightfield (BF) and are not collected.

25. Single fixed amoebae have an area of around 100–200 μm and an aspect ratio of just below one (i.e., a ratio of diameters; this equals one in a round object).

26. The Intensity versus Raw Max Pixel features are used to adjust the power of the lasers.

27. A sample containing all fluorophores should always be loaded first. This sample is used to adjust the power of the lasers. If the fluorescence at a certain fluorescence channel is expected to vary between samples, the brightest sample should be loaded first.

28. During acquisition, the gates can be set to define what is going to be collected. Thus, gates can be rather generous at this point, as it is important not to lose any events of interest. Later, at the analysis stage, the gates can be fine-tuned. Meanwhile, obviously unwanted objects (debris) should not be collected.

29. The Raw Max Pixel value (i.e., the intensity of the brightest pixel in the image) should be well above the background ($\approx$50) but below saturation level in each channel used.

30. The mPlum channel should be used for the identification events in focus for homogenized host cells, as a crisp bacterial image is crucial.

31. The final R2 gate should include only events clearly positive for both the *L. pneumophila* marker mPlum and the LCV marker P4C$_{SidC}$-mCherry, as these two signals define an LCV. The gate will be adjusted in later steps after a compensation matrix has been applied, as mCherry and mPlum have significant spectral overlap. It is important that the final gate is not too generous, leading to the acquisition of a high proportion of unwanted objects and resulting in insufficient numbers of remaining LCVs after gating.

32. Squelch is used to reduce the sensitivity of object detection. Host cell homogenization results in great amounts of debris, giving an abnormally high event rate. Squelching of debris by increasing the squelch value is necessary if the rate of total objects per second reaches 4000. Aim for a rate below 500 objects per second.

33. The first sample needs to be measured again after the laser powers have been adjusted. It is necessary to collect and apply a compensation matrix before proceeding to adjust the gates and acquire the experimental sample.

34. The Compensation Wizard will automatically turn off the brightfield and SSC illuminators and prompt you to acquire compensation control samples. For future runs with the same fluorophores (and identical laser powers), the acquisition settings, including the compensation matrix, can be saved as a template.

35. At this stage, it is helpful to rename the channels according to what they represent (e.g., BF for brightfield in Ch01, Cnx for calnexin in Ch02, P4C for P4C$_{SidC}$ in Ch04, Lpn for

L. pneumophila in Ch05), and to set the desired colors for each channel.

36. For the analysis of intact cells, wizards in IDEAS software, such as the Internalization Wizard, can be conveniently used. Select Ch05 (mPlum) as the internalizing probe and Ch01 (bright-field) as the cell image. Gate on the single-focused cells. The next step defines subpopulations based on the expression of specific probes. Select "Yes" and select the two markers used (Ch04 and Ch02; mCherry and GFP, respectively). It is important to exclude cells not expressing fluorescently tagged proteins to avoid false-positive colocalization results. Furthermore, gate cells that are associated with mPlum-positive bacteria, based on the Intensity and Max Pixel features in Ch05, and cells that have internalized the bacteria. In events with a high internalization value, the mPlum signal overlaps the image of the cell, indicating that the bacteria are inside the cell.

37. This step is included in order to select only cells carrying a single bacterium. Since IFC analyses events on a cell level (rather than each individual bacterium inside the cells), this step is important to enabling the classification of LCVs as "positive" or "negative" for a marker. If this classification is not required, host cells containing any number of internalized bacteria identified in the "Internalization Wizard" can be substituted in the remaining analysis steps.

    The resulting amoeba population, having internalized *L. pneumophila*, can be used directly in the Spot Wizard, with the first four steps skipped. In Ch05 (mPlum) assign two populations, one with a low spot count (i.e., only one bacterium per cell), and the other with a high spot count. Use the tagging tools for creating the image galleries. Make sure to include cells carrying bacteria of different intensities, shapes, and distances (approx. 30 cells in each population). The software will calculate the optimal mask for identifying the individual cells. Finally, a histogram showing the Spot Count feature in the resulting mask is created. It is noteworthy that this feature can also be used to determine the number of bacteria per cell, which can be useful for quantifying the intracellular replication of a pathogen [49]. For *L. pneumophila* in intact cells, we use the feature [Spot Count_Spot (M05, Lpn, Bright, 4, 1)].

38. Start with, e.g., mPlum and mCherry (Ch05 and Ch04). The amoeba population, having internalized only one bacterium, can be used in the Colocalization Wizard, with the first steps skipped. Finally, a histogram of the Bright Detail Similarity R3 feature for the cells containing only one bacterium is created. The Bright Detail Similarity R3 feature is the log-transformed

Pearson's correlation coefficient of the localized bright spots with a radius of 3 pixels or less within the masked area in the two input images. Thus, it represents the colocalization of the mCherry-tagged LCV marker and the mPlum-producing bacteria.

39. A gate can optionally be set to include only events where the bacteria colocalize with the marker (between 0.5–1.0). Use a biological negative control to confirm the gate. Quantitate the percentage within this gate or a measure of the central tendency of the Bright Detail Similarity R3 value.

40. The statistics report can include the following parameters: 1. Percentage of (FP (fluorescent proteins)-positive) cells having internalized bacteria ("% phagocytosis"). 2. The mean, median, and standard deviation of the Bright Detail Similarity R3 feature for mPlum/mCherry and mPlum/GFP, respectively, in the population having internalized only one bacterium. 3. The number of events in the population having internalized exactly one bacterium ("n"). 4. The percentage of cells the LCV positive for GFP or mCherry, respectively (i.e., the gate includes only events where the bacteria colocalize with the marker).

41. The compensation matrix applied during acquisition is not automatically applied during analysis. It is helpful to create a more fine-tuned compensation matrix at this stage.

42. Analysis of host cell homogenates is performed manually in IDEAS software since there are no suitable wizards for this purpose.

43. Only LCVs containing one bacterium are analyzed using the described methodology, as it is, in practice, very difficult to correctly discriminate multiple individual LCVs from LCVs containing multiple bacteria in an image. Thus, the method is more suitable for early time points post-infection, with few bacteria per LCV. The population identified in step 3 can be used directly in the Spot Wizard, and the first four steps of the wizard can be skipped. Next, select the spot image channel (Ch05 for mPlum). Then, assign two populations, one with a high spot count (i.e., many bacteria are visible in the image) and one with a low spot count (i.e., only one bacterium is visible). Use the tagging tools. Make sure to include bacteria of different intensities, shapes, and distances among them (approx. 30 events in each population). The software will calculate the optimal mask for identifying the individual bacteria. Finally, a histogram showing the Spot Count feature in the resulting mask is created, in which the cells having a spot count of one can be gated. For *L. pneumophila* in homogenates, we use the feature [Spot Count_Spot(M05, Lpn, Bright, 6, 1)].

44. First, use the Masks tool to create the mask [Object(M05, Lpn, Tight)], and then the Features tool to create the feature [Area_Object(M05, Lpn, Tight)].

45. First, use the Features tool to create the feature [Bright Detail Similarity R3_MC_Lpn_P4C].

46. It is important to use the Mean Pixel feature rather than the Intensity feature to quantify the GFP on the LCV. The Intensity feature represents the total intensity from all pixels in the image and will thus be biased by the size of the LCV, which is not the case for the Mean Pixel value, which represents the mean value of all the pixels in the mask.

47. The gate-defining calnexin-GFP-positive LCVs can be created directly on the already existing histogram by using a biological negative control. It can be further improved by creating an additional dot plot with [Mean Pixel_M02_Cnx] on the x-axis and [Bright Detail Similarity R3_MC_Lpn_Cnx] on the y-axis (this feature needs to be created first using the Features tool). Gate events with a high mean GFP signal (positive for calnexin) and with a high enough colocalization signal confirm that the GFP is located on the LCV.

48. The statistics report can include the following: 1. The mean, median, and standard deviation of the [Mean Pixel_MC_Cnx] feature in the *L. pneumophila*-containing and P4C$_{SidC}$-mCherry-positive LCVs (identified in step 5). 2. The number of events in the LCV population ("n"). 3. The percentage of LCVs positive for calnexin-GFP (i.e., contained in the gate created in Step 7).

Acknowledgments

Research in the laboratory of H.H. was supported by the Swiss National Science Foundation (SNF; 31003A_153200, 31003A_175557), the Novartis Foundation for Medical-Biological Research, the OPO foundation, and the Center of Microscopy and Image Analysis, University of Zürich (UZH). A.W. was supported by the Swedish Society of Medicine, the Linköping Society of Medicine, the Medical Inflammation and Infection Centre at Linköping University, and the Åke Wiberg Foundation. The funders had no role in study design, data collection and analysis, decision to publish, or preparation of the manuscript. Imaging flow cytometry was performed using equipment of the Flow Cytometry Unit, Core Facility, Medical Faculty, Linköping University.

References

1. Parte AC (2018) LPSN - list of prokaryotic names with standing in nomenclature (bacterio.net), 20 years on. Int J Syst Evol Microbiol 68:1825–1829

2. Hoffmann C, Harrison CF, Hilbi H (2014) The natural alternative: protozoa as cellular models for *legionella* infection. Cell Microbiol 16:15–26

3. Newton HJ, Ang DK, van Driel IR, Hartland EL (2010) Molecular pathogenesis of infections caused by *legionella pneumophila*. Clin Microbiol Rev 23:274–298

4. Swart AL, Harrison CF, Eichinger L, Steinert M, Hilbi H (2018) *Acanthamoeba* and *Dictyostelium* as cellular models for *legionella* infection. Front Cell Infect Microbiol 8: 61

5. Isberg RR, O'Connor TJ, Heidtman M (2009) The *legionella pneumophila* replication vacuole: making a cosy niche inside host cells. Nat Rev Microbiol 7:13–24

6. Hubber A, Roy CR (2010) Modulation of host cell function by *legionella pneumophila* type IV effectors. Ann Rev Cell Dev Biol 26:261–283

7. Asrat S, de Jesus DA, Hempstead AD, Ramabhadran V, Isberg RR (2014) Bacterial pathogen manipulation of host membrane trafficking. Ann Rev Cell Dev Biol 30:79–109

8. Finsel I, Hilbi H (2015) Formation of a pathogen vacuole according to *legionella pneumophila*: how to kill one bird with many stones. Cell Microbiol 17:935–950

9. Steiner B, Weber S, Hilbi H (2018) Formation of the *legionella*-containing vacuole: phosphoinositide conversion, GTPase modulation and ER dynamics. Int J Med Microbiol 308:49–57

10. Qiu J, Luo ZQ (2017) *Legionella* and *Coxiella* effectors: strength in diversity and activity. Nat Rev Microbiol 15:591–605

11. Bärlocher K, Welin A, Hilbi H (2017) Formation of the *legionella* replicative compartment at the crossroads of retrograde trafficking. Front Cell Infect Microbiol 7:482

12. Personnic N, Bärlocher K, Finsel I, Hilbi H (2016) Subversion of retrograde trafficking by translocated pathogen effectors. Trends Microbiol 24:450–462

13. Steiner B, Weber S, Kaech A, Ziegler U, Hilbi H (2018) The large GTPase atlastin controls ER remodeling around a pathogen vacuole. Commun Integr Biol 11:1–5

14. Urwyler S, Nyfeler Y, Ragaz C, Lee H, Mueller LN, Aebersold R et al (2009) Proteome analysis of *legionella* vacuoles purified by magnetic immunoseparation reveals secretory and endosomal GTPases. Traffic 10:76–87

15. Hoffmann C, Finsel I, Otto A, Pfaffinger G, Rothmeier E, Hecker M et al (2014) Functional analysis of novel Rab GTPases identified in the proteome of purified *legionella*-containing vacuoles from macrophages. Cell Microbiol 16:1034–1052

16. Naujoks J, Tabeling C, Dill BD, Hoffmann C, Brown AS, Kunze M et al (2016) IFNs modify the proteome of *legionella*-containing vacuoles and restrict infection via IRG1-derived itaconic acid. PLoS Pathog 12:e1005408

17. Schmölders J, Manske C, Otto A, Hoffmann C, Steiner B, Welin A et al (2017) Comparative proteomics of purified pathogen vacuoles correlates intracellular replication of *legionella pneumophila* with the small GTPase Ras-related protein 1 (Rap1). Mol Cell Proteomics 16:622–641

18. Bärlocher K, Hutter CAJ, Swart AL, Steiner B, Welin A, Hohl M et al (2017) Structural insights into *legionella* RidL-Vps29 retromer subunit interaction reveal displacement of the regulator TBC1D5. Nat Commun 8:1543

19. Steiner B, Swart AL, Welin A, Weber S, Personnic N, Kaech A et al (2017) ER remodeling by the large GTPase atlastin promotes vacuolar growth of *legionella pneumophila*. EMBO Rep 18:1817–1836

20. Welin A, Weber S, Hilbi H (2018) Quantitative imaging flow cytometry of *legionella*-infected *Dictyostelium* amoebae reveals the impact of retrograde trafficking on pathogen vacuole composition. Appl Environ Microbiol 84: e00158–e00118

21. Bozzaro S, Bucci C, Steinert M (2008) Phagocytosis and host-pathogen interactions in *Dictyostelium* with a look at macrophages. Int Rev Cell Mol Biol 271:253–300

22. Steinert M (2011) Pathogen-host interactions in *Dictyostelium, legionella, mycobacterium* and other pathogens. Stem Cell Dev Biol 22:70–76

23. Swart AL, Hilbi H (2020) Phosphoinositides and the fate of *legionella* in phagocytes. Front Immunol 11:25

24. Weber SS, Ragaz C, Reus K, Nyfeler Y, Hilbi H (2006) *Legionella pneumophila* exploits PI(4)P to anchor secreted effector proteins to the replicative vacuole. PLoS Pathog 2:e46

25. Weber SS, Ragaz C, Hilbi H (2009) The inositol polyphosphate 5-phosphatase OCRL1 restricts intracellular growth of *legionella*, localizes to the replicative vacuole and binds to the bacterial effector LpnE. Cell Microbiol 11: 442–460

26. Buckley CM, Heath VL, Gueho A, Bosmani C, Knobloch P, Sikakana P et al (2019) PIKfyve/

Fab1 is required for efficient V-ATPase and hydrolase delivery to phagosomes, phagosomal killing, and restriction of *legionella* infection. PLoS Pathog 15:e1007551

27. Peracino B, Balest A, Bozzaro S (2010) Phosphoinositides differentially regulate bacterial uptake and Nramp1-induced resistance to *legionella* infection in *Dictyostelium*. J Cell Sci 123:4039–4051

28. Hüsler D, Steiner B, Welin A, Striednig B, Swart AL, Molle V, Hilbi H, Letourneur F (2021) *Dictyostelium* lacking the single atlastin homolog Sey1 shows aberrant ER architecture, proteolytic processes and expansion of the *legionella*-containing vacuole. Cell Microbiol 23:e13318

29. Finsel I, Ragaz C, Hoffmann C, Harrison CF, Weber S, van Rahden VA et al (2013) The *legionella* effector RidL inhibits retrograde trafficking to promote intracellular replication. Cell Host Microbe 14:38–50

30. Rothmeier E, Pfaffinger G, Hoffmann C, Harrison CF, Grabmayr H, Repnik U et al (2013) Activation of ran GTPase by a *legionella* effector promotes microtubule polymerization, pathogen vacuole motility and infection. PLoS Pathog 9:e1003598

31. Swart AL, Gomez-Valero L, Buchrieser C, Hilbi H (2020) Evolution and function of bacterial RCC1 repeat effectors. Cell Microbiol 22:e13246

32. Veltman DM, Van Haastert PJ (2013) Extrachromosomal inducible expression. Methods Mol Biol 983:269–281

33. Weber S, Wagner M, Hilbi H (2014) Live-cell imaging of phosphoinositide dynamics and membrane architecture during *legionella* infection. MBio 5:e00839–13

34. Weber S, Steiner B, Welin A, Hilbi H (2018) *Legionella*-containing vacuoles capture PtdIns(4)*P*-rich vesicles derived from the Golgi apparatus. MBio 9:e02420–18

35. Welin A, Weber S, Hilbi H (2019) Quantitative imaging flow cytometry of *legionella*-containing vacuoles in dually fluorescence-labeled *Dictyostelium*. Methods Mol Biol 1921:161–177

36. Ragaz C, Pietsch H, Urwyler S, Tiaden A, Weber SS, Hilbi H (2008) The *legionella pneumophila* phosphatidylinositol-4 phosphate-binding type IV substrate SidC recruits endoplasmic reticulum vesicles to a replication-permissive vacuole. Cell Microbiol 10:2416–2433

37. Derre I, Isberg RR (2004) *Legionella pneumophila* replication vacuole formation involves rapid recruitment of proteins of the early secretory system. Infect Immun 72:3048–3053

38. Lu H, Clarke M (2005) Dynamic properties of *legionella*-containing phagosomes in *Dictyostelium* amoebae. Cell Microbiol 7:995–1007

39. Koliwer-Brandl H, Knobloch P, Barisch C, Welin A, Hanna N, Soldati T et al (2019) Distinct *Mycobacterium marinum* phosphatases determine pathogen vacuole phosphoinositide pattern, phagosome maturation, and escape to the cytosol. Cell Microbiol 21:e13008

40. Sadosky AB, Wiater LA, Shuman HA (1993) Identification of *legionella pneumophila* genes required for growth within and killing of human macrophages. Infect Immun 61:5361–5373

41. Segal G, Purcell M, Shuman HA (1998) Host cell killing and bacterial conjugation require overlapping sets of genes within a 22-kb region of the *legionella pneumophila* genome. Proc Natl Acad Sci U S A 95:1669–1674

42. Horwitz MA (1983) The Legionnaires' disease bacterium (*legionella pneumophila*) inhibits lysosome-phagosome fusion in human monocytes. J Exp Med 158:2108–2126

43. Feeley JC, Gibson RJ, Gorman GW, Langford NC, Rasheed JK, Mackel DC et al (1979) Charcoal-yeast extract agar: primary isolation medium for *legionella pneumophila*. J Clin Microbiol 10:437–441

44. Loovers HM, Kortholt A, de Groote H, Whitty L, Nussbaum RL, van Haastert PJ (2007) Regulation of phagocytosis in *Dictyostelium* by the inositol 5-phosphatase OCRL homolog Dd5P4. Traffic 8:618–628

45. Cocucci SM, Sussman M (1970) RNA in cytoplasmic and nuclear fractions of cellular slime mold amebas. J Cell Biol 45:399–407

46. Malchow D, Nagele B, Schwarz H, Gerisch G (1972) Membrane-bound cyclic AMP phosphodiesterase in chemotactically responding cells of *Dictyostelium discoideum*. Eur J Biochem / FEBS 28:136–142

47. Barisch C, Paschke P, Hagedorn M, Maniak M, Soldati T (2015) Lipid droplet dynamics at early stages of *Mycobacterium marinum* infection in *Dictyostelium*. Cell Microbiol 17:1332–1349

48. Veltman DM, Akar G, Bosgraaf L, Van Haastert PJM (2009) A new set of small, extrachromosomal expression vectors for *Dictyostelium discoideum*. Plasmid 61:110–118

49. Johansson J, Karlsson A, Bylund J, Welin A (2015) Phagocyte interactions with *Mycobacterium tuberculosis*--Simultaneous analysis of phagocytosis, phagosome maturation and intracellular replication by imaging flow cytometry. J Immunol Methods 427:73–84

Chapter 5

Imaging Flow Cytometry of Multi-Nuclearity

Ivan A. Vorobjev, Sultan Bekbayev, Adil Temirgaliyev, Madina Tlegenova, and Natasha S. Barteneva

Abstract

Multi-nuclearity is a common feature for cells in different cancers. Also, analysis of multi-nuclearity in cultured cells is widely used for evaluating the toxicity of different drugs. Multi-nuclear cells in cancer and under drug treatments form from aberrations in cell division and/or cytokinesis. These cells are a hallmark of cancer progression, and the abundance of multi-nucleated cells often correlates with poor prognosis.

The use of standard bright field or fluorescent microscopy to analyze multi-nuclearity at the quantitative level is laborious and can suffer from user bias. Automated slide-scanning microscopy can eliminate scorer bias and improve data collection. However, this method has limitations, such as insufficient visibility of multiple nuclei in the cells attached to the substrate at low magnification.

Since quantification of multi-nuclear cells using microscopic methods might be difficult, imaging flow cytometry (IFC) is a method of choice for this. We describe the experimental protocol for the preparation of the samples of multi-nucleated cells from the attached cultures and the algorithm for the analysis of these cells by IFC. Images of multi-nucleated cells obtained after mitotic arrest induced by taxol, as well as cells obtained after cytokinesis blockade by cytochalasin D treatment, can be acquired at a maximal resolution of IFC. We suggest two algorithms for the discrimination of single-nucleus and multi-nucleated cells. The advantages and disadvantages of IFC analysis of multi-nuclear cells in comparison with microscopy are discussed.

Key words Mitosis, Microtubule inhibitors, Cytochalasin D, Multi-nuclearity

---

1 Introduction

Cancer is related to genetic processes such as genome instability, rearrangements, or specific gene mutations, followed by epigenetic changes, which ultimately lead to deregulated cell proliferation. One of the specific features of cancer is the presence of multi-nucleated cells [1, 2]. In normal tissues, multi-nucleated cells of macrophage origin like foreign body giant cells, Langerhans' cells, and osteoclasts are present [3]. Besides, multi-nucleated cells of fibroblast origin are formed in some pathological processes

Natasha S. Barteneva and Ivan A. Vorobjev (eds.), *Spectral and Imaging Cytometry: Methods and Protocols*,
Methods in Molecular Biology, vol. 2635, https://doi.org/10.1007/978-1-0716-3020-4_5, © The Author(s) 2023

[4, 5]. The most interesting, however, is the formation and dynamics of multi-nucleated cells of cancer origin.

Multi-nucleation is observed in many cancers [6–8]. The abundance of polyploid multi-nucleated cells (MNCs) strongly correlates with resistance to chemotherapy [8, 9]. Dormant large multi-nucleated cells correlate with poor prognosis and are a hallmark of relapse after anticancer treatment (reviewed in [1, 2]). The frequency of the multi-nucleated cells in cancers increases after different chemotherapy treatments with taxanes [10, 11], doxorubicin [8, 9, 12], carboplatin [9, 13–15], and other drugs.

Polyploid multi-nucleated cancer cells can be generated by several mechanisms: (i) mitosis without cytokinesis resulting in the formation of plasmodium-like cell; (ii) cell-cell fusion with the formation of the syncytium, and (iii) through mitotic arrest followed by exit into the interphase without chromosome separation (mitotic slippage). The outcome of these processes is different. The first two mechanisms result in the formation of giant cells, with each nucleus of nearly normal size. In the last case, the nuclei formed are numerous and might be rather small since they contain one or few chromosomes [16]. These are called micronuclei. The technical challenge is whether one can distinguish such micronuclei from the lobulated ones formed by other processes [17] and also present in the cancer cell populations.

Currently, quantitative analysis of the multi-nucleated cells is hampered because of the limitations of manual assessment of the multi-nucleated cell frequency. Multi-nucleated cells containing micronuclei could be evaluated under the microscope only using relatively high magnification. This requires prolonged observations with 3D visualization; otherwise, results might be biased. The manual microscopy-based analysis is labor-intensive and time-consuming. Automated microscopy can partially eliminate this bias but suffers from the lack of image processing algorithms [18]. The task of cell nuclei counting is challenging due to the required nuclear segmentation of overlapping and/or touching nuclei and the presence of noise and image acquisition variables [19, 20]. Abnormalities in nuclei texture or shape are hallmarks of cancer cells, and highly textured nuclei fluorescent images can lead to apparent undesirable splitting of a single nucleus during image segmentation.

Of particular interest is the formation of multi-nucleated cells after mitotic arrest induced by anti-microtubule drugs frequently happening in cancer cells [21]. Microscopic evaluation of the frequency of mitotic slippage requires prolonged time-lapse recordings and manual data analysis of numerous time-lapse series [22–24]. Besides, many cells die soon after slippage [23, 25]. Thus, determining the outcome of mitotic arrest induced by drug treatments in cancer cells remains challenging.

Imaging flow cytometry (IFC) is a method of choice in the morphometric research of highly heterogeneous populations of cells [26]. IFC advantages are (1) imaging of single cells excluding a requirement of cellular segmentation and (2) high-throughput capabilities of instrumentation, allowing a standardized analysis of tenths to thousands of images based on morphological and fluorescent features.

Among the different types of nuclei, there are three nuclei types of particular interest: (1) lymphoid cell nuclei that are of regular shape and relatively small size; (2) epithelial cell nuclei with homogeneous, nearly uniform chromatin distribution; (3) nuclei of high-grade cancer cells that have irregular, pleomorphic boundaries, clear nucleoli and heterogeneous chromatin [27].

A detailed description of IFC application for the development of micronucleus assay using cells of lymphoid origin was given by Rodrigues and co-authors [18, 28], Verma and co-authors [29], and recently was expanded to skin epithelial cells [30]. The procedure of multi-nuclearity assessment might combine several basic image analysis algorithms such as watershed-based nuclei segmentation, thresholding, and intensity-based spot-counting. Watershed transform is one of the most popular approaches for nuclear and cellular image segmentation [31–33]. Spot-counting algorithms were intensively used for FISH and micronucleus image counting in gene toxicity and radiotoxicity research [34]. Machine-learning and deep learning approaches were developed recently for the analysis of complex label-free and immunofluorescent images where simple segmentation is not sufficient [35, 36].

Multi-nuclearity evaluation, as well as other nuclear morphometric and image features such as image texture, nucleus-to-cytoplasm ratio, nucleus size, pleomorphism degree, can be helpful in the evaluation of cancer grade and treatment efficiency [37]. Image analysis based on the texture parameters, i.e., radiomics and not on morphometry (length, area, size), remains challenging even after successful nuclear segmentation [38–40] and quantitation of cancer cells types due to the heterogeneous chromatin distribution in their nuclei still not possible. In the present study, we extend the IFC application for the analysis of attached cultured cancer cells with pleomorphic nuclei and show that promising results could be obtained by using the developed protocol.

Addressing the question about the feasibility of IFC to analyze the formation of multi-nucleated cells in cancer A549 cells (non-small lung cancer carcinoma), we tested two models – cytochalasin D-induced multi-nucleation and multi-nucleation after mitotic slippage under paclitaxel treatment after Hoechst 33342 staining. In the first model, the formation of multi-nucleated cells occurs through "endomitosis", i.e., mitosis without cytokinesis [41]. According to the microscopic analysis, these cells usually have 3–6 nuclei of the same size and form [42]. In the second model, multi-nucleated cells are tetraploid ones that have formed after mitotic slippage [24]. These cells contain a different number of nuclei, including so-called micronuclei [43].

Galleries of multi-nucleated cells and cells with a single nucleus can be easily obtained (Fig. 1a and b) manually; however, many images come into the "gray" zone. The use of the EDF (extended depth of focus) function resulted in a significant enhancement of the contrast of the fluorescent images making the texture more evident but precluded visual analysis of multi-nuclearity.

The percentage of cells with more than one nucleus was further determined using machine-learning approaches. The standard approach to analyze multi-nuclearity described in detail by Rodriguez and others [18, 28] has limitations in the discrimination of mononucleated and multi-nucleated cells with pleomorphic nuclei because of the insufficient gradients at the nuclear edges. When EDF option was activated, nuclear staining became more heterogeneous, and also analysis by a complex of masks was inefficient. Since some of the multi-nucleated cells could be easily distinguished by manual visual inspection, we further employed a machine-learning algorithm. This protocol gave a good separation of the test galleries (Fig. 2a).

Statistical data were collected for both categories (single-nucleated and multi-nucleated) and were comparable with the results obtained by machine-learning algorithm (multi-classifier 2 at Fig. 2b). For the comparison we took a useful parameter that is cell diameter/area (cell volume). The area size of multi-nucleated taxol-treated cells was 118% from the single-nucleated taxol-treated cell population and the overlap between two distributions is larger than for the multi-classifier (Fig. 2c).

Manual inspection of image galleries obtained from control and taxol-treated cells using different cut-off values on the multi-classifier axis show that undoubtedly single nucleated cells are located on the multi-classifier histogram below -0.3, and undoubtedly multi-nucleated cells are located on the histogram above 0.5 (Fig. 3d and e). We suggest that a reasonable estimate of the minimal percentage of multi-nucleated cells with a high degree of heterogeneity in chromatin staining can be obtained using a machine-learning approach.

For a more detailed analysis of multi-nuclear cells, we used high-resolution confocal microscopy. Confocal microscopy taken with an equivalent pixel size of 70 nm in X-Y plane and step of 100 nm along the Z axis (LSM 780, PlanApo 63×/1.4 Oil immersion objective) shows complex morphology of nuclei in untreated cells (Fig. 3a) and gives a larger number of nuclei after treatments (Fig. 3b), but takes a lot of time for analysis of individual cells (scanning of one z-stack across the cell takes about 5–10 min) and does not allow evaluation of the frequency of multi-nucleated cells in the population. However, in some cases in the cells that have

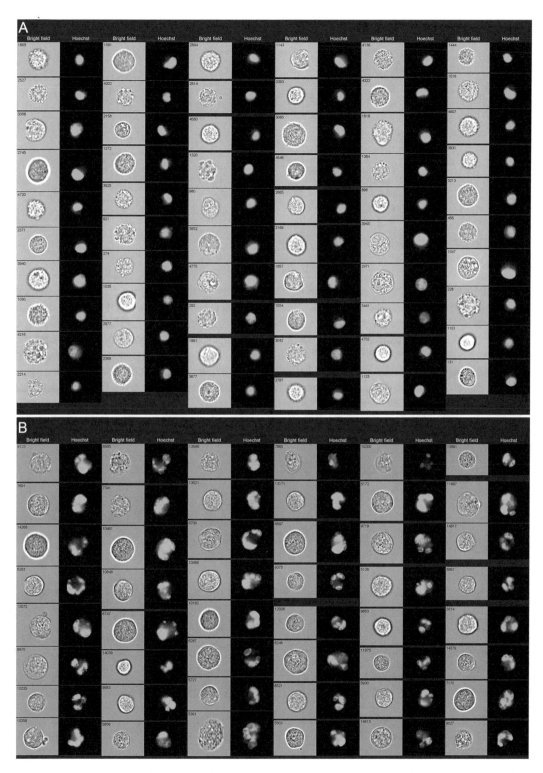

Fig. 1 (**a**) Mono-nucleated A549 cells. (**b**) Multinucleated A549 cells collected after mitotic slippage. The presence of multiple nuclei/nuclear lobes is evident, but discrimination between multilobularity (one nucleus with many lobes) and true multi-nuclearity (many nuclei) is not possible. Notice that the size difference between cells in (**a**) and (**b**) is negligible (Brightfield)

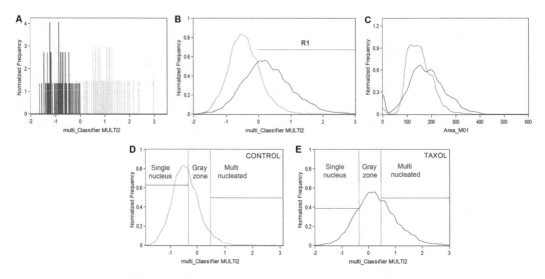

Fig. 2 (**a**) Histograms of mono- and multi-nucleated cells from training galleries for machine learning algorithm (purple- mono-nucleated and yellow – multi-nucleated cells). (**b**) Histograms of events distribution along Classifier Multi2. (Green – control A549 cells, red – A549 cells treated with taxol (analyzed with FlowJo (BD Biosciences, USA). R1 defines the region of multi-nucleated cells. (**c**) Histograms of Area parameter in control and treated with taxol A549 cells (128% of area) (blue – mono-nucleated cells; red- multi-nucleated cells). (**d-e**) Histograms of Control (Green) and Taxol-treated (Red) cellular populations distributed along multi-classifier 2. Gated mono-nucleated and multi-nucleated regions in each cell population, and Gray Zone (between −0.3 and 0.5 scores) where identification of multi-nuclearity is ambiguous.

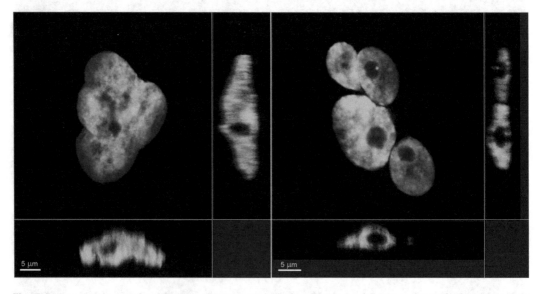

Fig. 3 Laser scanning confocal images of mono-nucleated multi-lobular (**a**) and multi-nucleated (**b**) cells in culture. X-Y projection is given on the right; X-Z projection is given on the left, Y-Z projection is given at the bottom. Hoechst 33342 staining. Scale bar 5 microns

formed after mitotic slippage, exact discrimination of multiple nuclei from each other is not possible even by microscopy. Thus, the detailed analysis of the multi-nuclearity complex approach - IFC for the frequency of multi-nucleated cells and confocal microscopy for the estimation of the average number of nuclei per cell - is recommended.

We conclude that though IFC is limited in recognition of multiple micronuclei from the multi-lobed mononuclear cancer cells with pleomorphic nuclei, it stays the best method to determine the frequency of multi-nucleated cancer cells.

2 Materials

2.1 Laboratory Instrumentation and Accessories

1. Imagestream X MKII (Amnis-Luminex, USA) 4-lasers system with 60× magnification.

2. Laser scanning confocal microscope (LSM-780, Zeiss, Germany).

3. Centrifuge without refrigeration (Eppendorf Centrifuge 5702, USA).

4. Light inverted microscope (Zeiss Primovert, Germany).

5. Motorized inverted fluorescent microscope (Zeiss Cell Observer, Germany).

6. Thermoblock (Thermo Fisher Scientific 3121 Water-Jacketed CO_2 Incubator with IR Sensor, USA).

7. Standard laminar flow hood (Thermo Fisher Scientific, USA).

8. Standard hemocytometer.

9. Sterile cultural ventilated flasks ($25\ cm^2$) and multi-well plates (TPP, Switzerland).

10. 15 ml centrifugation tubes (TPP, Switzerland).

11. Petri dishes with a glass-bottom (Corning, USA).

2.2 Reagents

1. Complete DMEM medium: DMEM supplemented with L-glutamine, Pen-Strep (penicillin-streptavidin), and 5% fetal bovine serum (FBS).

2. CO_2-independent DMEM medium (Gibco™, ThermoFisher 18045088, USA).

3. 0.5% CO_2-independent DMEM media with 1% FBS (DMEM with 1% FBS was diluted 1:1 by distilled water). For this, 25 mL of sterile distilled water was mixed thoroughly with 25 mL CO_2-independent medium in laminar-flow hood. 500 μL was removed from this solution, and 500 μL of FBS was added and mixed thoroughly.

4. Fetal Bovine Serum (Sigma F2442, USA, sterile-filtered, suitable for hybridoma).

5. DMEM media, penicillin, streptomycin, L-glutamine suitable for cell culture.

6. Hoechst 33342.

7. Taxol (Paclitaxel, Sigma T7402, USA).

8. Cytochalasin D.

9. Trypsin-EDTA 0.25% sterile solution.

10. Phosphate buffered saline.

3   Methods

3.1   Cancer Cell Line A549 Culturing

1. Grow A549 cancer cell line in DMEM complete culture medium supplemented with L-glutamine, Pen-Strep. and 5% FBS in 25 cm^2 culture flasks or multi-well plates (6-well plates (TPP, Switzerland)).

3.2   Growing A549 Cellular Monolayer with Taxol

1. After cells reach a subconfluent density, add Taxol (Sigma, USA) to the final concentration of 100 or 300 ng/mL (100–300 nM) and leave for 24 h in the CO_2 incubator.

2. After 24 h, remove the Taxol by changing the culture medium.

3. Wash a monolayer with 5 ml of the culture medium and gently shake the flask.

4. Transfer cultural medium with detached mitotic cells into a 15 mL tube.

5. Centrifuge mitotic cells in cell centrifuge at 300 g for 3 min.

6. After centrifugation, dispose a supernatant and collect pellet in 7 mL of complete DMEM with 5% FBS.

7. Centrifuge cell suspension again in fresh culture medium at 300 g for 3 min.

8. Dispose supernatant again, resuspend a pellet carefully in 7 ml of a complete DMEM, and seed cells in a new Petri dish (*see* **Note 1**).

9. Incubate cells in a Petri dish in a drug-free medium in CO_2 incubator at 37 °C and 5% CO_2 for 12 h.

3.3   Growing A549 Cellular Monolayer with Cytochalasin D

1. After cells reach a subconfluent density in multi-well plates add Cytochalasin D (Sigma, USA) to a final concentration of 0.5 μM and culture cells with the drug for 72 h.

2. After three days of incubation, remove the drug by replacing three times the culture medium with drug-free complete DMEM with 5% FBS.

3. Incubate cells in a drug-free medium in CO_2 incubator at 37 °C and 5% CO_2 for another 24 h.

3.4   Preparation of Multi-Nucleated Cellular Suspension after Taxol or Cytochalasin D Treatment for Analysis on Imagestream X Mark II

1. After 12 h of post-Taxol growing of A549 cellular monolayer in Petri dish, rinse cells three times with 3 mL 1× PBS.

2. Then rinse cellular monolayer two times with 2 mL solution of 0.05% Trypsin with 0.5 mM EDTA (in PBS).

3. Incubate cells in a Petri dish in 1 mL solution of Trypsin (0.05%) with 0.5 mM EDTA for 3 min (*see* **Note 2**).

4. After detach cells from the dish surface, add 5 mL DMEM with 5% FBS, resuspend cells, and collect in 15 mL centrifuge tube.

5. Take 10 μL of cell suspension and count cells with a standard hemocytometer or automatic hemocytometer Countess 3 (Thermo Fisher Scientific, USA).

6. Prepare cellular suspension containing 500,000 cells and centrifuge at 300 g for 3 min.

7. After centrifugation, dispose a supernatant and resuspend a cellular pellet in 2 mL of hypotonic (0.5×) CO_2-independent medium containing 1% FBS with 5.5 μM Hoechst 33342.

8. Transfer cell suspension in the Eppendorf tube and incubate in a thermostat at 37 °C for 30 min with periodical agitation every 5 min.

9. At the end of the incubation centrifuge a cellular suspension in Eppendorf centrifuge at 300 g for 3 min.

10. Dispose the major part of the supernatant, and resuspend pellet collecting 100,000–500,000 cells in approximately 60 μL of remaining supernatant.

3.5   Analysis of Micronuclei in Multi-Nucleated Cells under Microscope

1. Seed a cellular suspension of A549 cell line after treatment with Taxol for 24 h and subsequential drug removal in 3.5 cm Petri dishes with a glass bottom (Corning, USA).

2. Incubate cells in a Petri dish in CO_2-incubator 12 h at 37 °C 5% CO_2.

3. After 12 h of incubation, dispose cultural medium and rinse cells with 2 mL 1× PBS pre-warmed up to 37 °C.

4. Add 2 mL of 1% glutaraldehyde solution in 1× PBS to the cells and incubate for 30–60 min (*see* **Note 3**).

5. Remove a glutaraldehyde solution from a Petri dish.

6. Rinse cell monolayer in the Petri dish with 1×PBS for 30 min by changing 1×PBS three times every 10 min.

7. Add 1× PBS containing 5.5 μM Hoechst 33342 to a Petri dish for at least for 30 min (*see* **Note 4**).

8. Take images using Zeiss Axio Observer microscope (Zeiss, Germany) equipped with 63×/1.46 oil immersion objective and Hamamatsu ORCA-FLASH II camera in DIC and DAPI channels as 5×5 tiles with z-stack (15 μm size using 0.3 μm step).

9. Deconvolve z-stacks using Huygens software (Scientific Volume Imaging, Netherlands) and count nuclei manually.

3.6 Preparation Cells for Imaging Flow Cytometry Analysis

1. Seed A549 cells line after treatment with Taxol for 24 h and drug removal into 3.5 cm Petri dishes (Corning, USA).

2. Incubate cells in a Petri dish in CO_2-incubator for 12 h at 37 °C 5% CO_2.

3. After 12 h incubation of taxol-treated cells in the drug-free medium, dispose culture medium and rinse cells 3 times with 3 mL of PBS.

4. Then rinse cells 2 times with 2 mL Trypsin 0.05% with 0.5 mM EDTA, and incubate in 1 mL Trypsin (0.05%) with 0.5 mM EDTA for 3 min.

5. After cells detach from the surface, add 5 mL DMEM with 5% FBS, resuspend cells, and collect in 15 mL tube (*see* **Note 5**).

6. Dispose the supernatant and resuspend the pellet of cells in 2 mL hypotonic (0.5×) CO_2-independent medium containing 1% FBS with 5.5 μM Hoechst 33342 (*see* **Note 6**).

7. After incubation in hypotonic medium for 10 min., sediment the cells by centrifugation at 300 g for 3 min.

8. Dispose the major part of the supernatant cautiously and resuspend the pellet in 60–100 μL of the same medium (*see* **Note 7**).

3.7 Starting and Running Imagestream X Mark II

1. Fill out instrument sheath and cleaning containers (cleanser, debubbler, sterilizer), empty waste, and check the speed beads container. Load **Default Template** from **File** Menu. Run all required calibrations and tests (*see* **Note 8**).

2. Select magnification at 60×, and high sensitivity mode. Turn on and optimize power of 405 nm (and 488 nm, if required) lasers. Disable other lasers (see **Note 9**).

3. Select the **Scatterplot** icon. Then select Area M01 on the X-axis and Aspect Ratio M1 on the Y-axis. Draw the region around single cells.

4. Set the acquisition parameters, specify the file name and the destination folder, and change the events number to 10,000-20,000. Collect bright-field imagery in channels 1 and 9, and side scatter in channel 6, minimizing signal and avoiding saturation.

5. Click on **Load** and apply an Eppendorf tube with sample in the sample port when prompted to do so.

6. Acquire 10,000–20,000 cellular events. Define nuclear events as positive for staining with Hoechst 33342. Create a histogram using as a parameter **Intensity M07** and draw a region defining cellular events positive for Hoechst 33342.

7. Launch the IDEAS software analysis package. Spectral compensation is not required unless additional staining is involved. Open the file in the IDEAS software.

8. Create analysis template for the identification of single events using dotplot with parameters **Area M01** on the X-axis and **Aspect Ratio M1** on the Y-axis, and create a region around single cells, and name it **Single Cells**.

9. Create a dotplot with Gradient **RMS M01** on the X-axis, and Gradient **RMS M07** on the Y-axis. Select **Single Cells** as a parent population. Draw the region and name it **Focused Cells**.

10. Create two galleries with manually picked cells with clearly identifiable single individual nuclei (Fig. 1) and with multi-nucleated cells (> 30 cells each).

3.8 Evaluation of Multi-Nuclear Cells Using Machine-Learning Module

In this protocol, a machine-learning module of "IDEAS" vs. 6.3 software (Amnis-Luminex, USA) is used.

1. After creating training galleries of images, start machine-learning Module to create a new parameter for image analysis.

2. Choose populations (single, focused) if the populations are **GATED** or **ALL** if they are not gated.

3. After populations of interest have been tagged, and galleries of images were chosen, use control-click to select training galleries (single and multi-nucleated cells). Alternatively, use the tagging tool to create galleries (*see* **Note 11**).

4. To exclude existing irrelevant features from different categories during step 3, select categories that best discriminate your data (*see* **Note 12**).

5. Choose channels to analyze the images (Channel 7 – Hoechst 33343 staining).

6. Click the "Start" button to create a single multi-parameter, and verify its content.

7. After creating the classifier, you can **Edit** or **Finish** and exit the wizard.

8. View new multi-classifier features in the feature manager. Click on the Analysis tab and select a new feature created and see the weight of different components in the new Feature.

9. Apply a new multi-component classifier to the populations of interest. For example, multi-classifier 1 (a combination of Mean and Min Pixel and Lobe Count features that uses Morphology and Skeleton Masks for Channel 7 (Hoechst 33343); multi-feature classifier 2 (a combination of Bright Detail Intensity and Variance Mean that uses fluorescent mask from Channel 7).

10. Create histograms from merged file to compare control and taxol-treated samples using a new chosen multi-feature.

11. Export FCS information from a file and analyze it using FlowJo (Treestar-BD Biosciences, USA) or FACSExpress (De Novo software, USA) off-line analysis program. Apply the Kolmogorov-Smirnov test to verify whether differences between overlapping histograms are significant (Fig. 2b).

4   Notes

1. If the multi-nucleated cells will be observed with a microscope, use glass-bottom Petri dishes (Corning, USA).

2. Time needed for the complete detachment of cells might vary. Incubation should be ended when a majority of cells start rolling on the surface of the flask.

3. Cells could be fixed with PFA; however, it will result in slight shrinkage of cells making analysis of the multi-nuclearity more difficult.

4. Images could be taken with laser scanning confocal microscope; however, this method is laborious and not useful for quantification of the proportion of multi-nucleated cells in the cellular population. This stage is important for the 3D reconstruction and determining the true number of micronuclei per cell, but not for the enumeration of the proportion of multi-nucleated cells.

5. This stage is important for the 3D reconstruction and for determining the number of micronuclei per cell, but not for the enumeration of multi-nucleated cells.

6. At this stage, 10 µL of suspension should be taken, and number of cells should be counted on a hemacytometer. Whether the overall concentration of cells exceeds 1 mln/ml, you can continue in the way described below. When concentration of cells is lower, you should cautiously collect cells after each step. When the concentration of cells is below 500,000 per ml, the number of cells after the end of specimen preparation could be low, and this will slow down ImageStream-based acquisition even when cells will be resuspended in the minimal volume.

7. Cell suspension in the Eppendorf tube should be agitated every 5–7 min and incubated in 37 °C for 30 min.

8. If the concentration of cells after first sedimentation (step 5) was low (less than 1 mln/mL), the volume for resuspension should be minimal – about 40–50 µL.

9. When you click "Startup" button make sure that the checkbox for Start all calibrations and tests is checked. It will calibrate the instrument following the start up.

10. Make sure that all cells can be seen in the Cell View column and that debris is excluded from viewing and acquisition (use the Single Cell gating for this purpose). Exclude the images with saturated pixels by changing the excitation laser powers.

11. The IDEAS software (Amnis-Luminex, USA) machine-learning module requires at least 25 images to be included in the training galleries. We recommend 40–50 images if it is possible. Unobvious and/or controversial images should not be included into training galleries. More than two galleries can be created and used.

12. Try to exclude different categories or some features irrelevant to the task during step 3 of creating multi-parameter classifier.

Acknowledgments

Work was supported by the Ministry of Health of the Republic of Kazakhstan under the program-targeted funding of the Ageing and Healthy Lifespan research program (IRN: 51760/Φ-М Р-19)) and AP14869915 (Ministry of Education and Science, Kazakhstan) to IAV. NSB was funded by CRP 16482715 and Faculty Development SSH2020028 grants from Nazarbayev University.

References

1. Mirzayans R, Andrais B, Murray D (2018) Roles of polyploid/multinucleated giant cancer cells in metastasis and disease relapse following anticancer treatment. Cancers 10:118–129. https://doi.org/10.3390/cancers10040118

2. Mirzayans R, Murray D (2020) Intratumor heterogeneity and therapy resistance: contributions of dormancy, apoptosis reversal (anastasis) and cell fusion to disease recurrence. Int J Mol Sci 21:1308. https://doi.org/10.3390/ijms21041308

3. Anderson JM (2000) Multinucleated giant cells. Curr Opin Hematol 7:40–47. https://doi.org/10.1097/00062752-200001000-00008

4. Regezi JA, Courtney RM, Kerr DA (1975) Fibrous lesions of the skin and mucous membranes which contain stellate and multinucleated cells. Oral Surg Oral Med and Oral Pathol 39:605–614

5. Cho MI, Garant PR (1984) Formation of multinucleated fibroblasts in the periodontal ligaments of old mice. Anat Rec 208:185–196

6. Fishback NF, Travis WD, Moran C, Guinee D Jr, McCarthy W, Koss M (1994) Pleomorphic (spindle/giant cell) carcinoma of the lung. Cancer 73:2936–2945

7. Ryska A, Reynolds C, Keeney GL (2001) Benign tumors of the breast with multinucleated stromal giant cells. Immunohistochemical analysis of six cases and review of the literature. Virchows Arch 439:768–775. https://doi.org/10.1007/s004280100470

8. Parekh A, Das S, Parida S, Das CK, Dutta D, Mallick SK (2018) Multi-nucleated cells use ROS to induce breast cancer chemo-resistance in vitro and in vivo. Oncogene 37:4546–4561. https://doi.org/10.1038/s41388-018-0272-6

9. Rohnalter V, Roth K, Finkernagel F, Adhikary T, Obert J, Dorzweiler K et al (2015) A multi-stage process including

transient polyploidization and EMT precedes the emergence of chemoresistant ovarian carcinoma cells with a dedifferentiated and proinflammatory secretory phenotype. Oncotarget 6:40005–40025. https://doi.org/10.18632/oncotarget.5552

10. Kadota K, Suzuki K, Colovos C, Sima CS, Rusch VW, Travis WD, Adusumilli PS (2012) A nuclear grading system is a strong predictor of survival in epithelioid diffuse malignant pleural mesothelioma. Mod Pathol 25:260–271. https://doi.org/10.1038/modpathol.2011.146

11. Ogdena A, Ridaa PCG, Knudsenb B, Kucukc O, Anejaa R (2015) Docetaxel-induced polyploidization may underlie chemoresistance and disease relapse. Cancer Lett 367:89–92. https://doi.org/10.1016/j.canlet.2015.06.025

12. Weihua Z, Lin Q, Ramoth AJ, Fan D, Fidler IJ (2011) Formation of solid tumors by a single multinucleated cancer cell. Cancer 117:4092–4099. https://doi.org/10.1002/cncr.26021

13. Bagnyukova TV, Serebriiskii IG, Zhou Y, Hopper-Borge EA, Golemis EA, Astasturov I (2010) Chemotherapy and signaling: how can targeted therapies supercharge cytotoxic agents? Cancer Biol Ther 10:839–853. https://doi.org/10.4161/cbt.10.9.13738

14. Kuo CH, Lu YC, Tseng YS, Shi CS, Chen SH, Chen PT, Wu FL, Chang YP, Lee YR (2014) Reversine induces cell cycle arrest, polyploidy, and apoptosis in human breast cancer cells. Breast Cancer 21:358–369. https://doi.org/10.1007/s12282-012-0400-

15. Lu YC, Lee YR, Liao JD, Lin CY, Chen YY, Chen PT et al (2016) Reversine induced multinucleated cells, cell apoptosis and autophagy in human non-small cell lung cancer cells. PLoS One 11:e0158587. https://doi.org/10.1371/journal.pone.0158587

16. Zhu Y, Zhou Y, Shi J (2014) Post-slippage multinucleation renders cytotoxic variation in anti-mitotic drugs that target the microtubules or mitotic spindle. Cell Cycle 13:1756–1764. https://doi.org/10.4161/cc.28672

17. Nakayama Y, Yamaguchi N (2005) Multilobulation of the nucleus in prolonged S phase by nuclear expression of Chk tyrosine kinase. Exp Cell Res 304:570–581. https://doi.org/10.1016/j.yexcr.2004.11.027

18. Rodrigues MA (2018) Automation of the in vitro micronucleus assay using the Image-Stream imaging flow cytometer. Cytometry A 93A:706–726. https://doi.org/10.1002/cyto.a.23493

19. Coelho LP, Shariff A, Murhy RF (2009) Nuclear segmentation in microscope cell images: a hand-segmented dataset and comparison of algorithms. In: IEEE international symposium on biomedical imaging: from Nano to macro. ISBI'09. IEEE, pp 518–521

20. Xing F, Yang L (2013) Robust nucleus/cell detection and segmentation in digital pathology and microscopy images: a comprehensive review. IEEE Rev Biomed Engineer 9:234–263

21. Gascoigne KE, Taylor SS (2009) How do antimitotic drugs kill cancer cells? J Cell Sci 122:2579–2585. https://doi.org/10.1242/jcs.039719

22. Gascoigne KE, Taylor SS (2008) Cancer cells display profound intra- and interline variation following prolonged exposure to antimitotic drugs. Cancer Cell 14:111–122. https://doi.org/10.1016/j.ccr.2008.07.002

23. Orth JD, Kohler RH, Foijer F, Sorger PK, Weissleder R, Mitchison TJ (2011) Analysis of mitosis and antimitotic drug responses in tumors by in vivo microscopy and single-cell pharmacodynamics. Cancer Res 71:4608–4616. https://doi.org/10.1158/0008-5472.CAN-11-0412

24. Cheng B, Crasta K (2017) Consequences of mitotic slippage for antimicrotubule drug therapy. Endocr Relat Cancer 24:T97–T106. https://doi.org/10.1530/ERC-17-0147

25. Shi J, Orth JD, Mitchison T (2008) Cell type variation in responses to antimitotic drugs that target microtubules and kinesin-5. Cancer Res 68:3269–3276

26. Barteneva N, Vorobjev IA (2018) Cellular heterogeneity: methods and protocols. Met Mol Bio 1745. https://doi.org/10.1007/978-1-4939-7680-56

27. Irshad H, Veillard A, Roux L, Racoceanu D (2014) Methods for nuclei detection, segmentation and classification in digital histopathology: a review-current status and futire potential. IEEE Rev Biomed Engineering 7:97–114

28. Rodrigues MA, Beaton-Green LA, Kutzner BC, Wilkins RC (2014) Automated analysis of the cytokinesis-block micronucleus assay for radiation biodosimetry using imaging flow cytometry. Radiat Environ Biophys 53:273–282. https://doi.org/10.1007/s00411-014-0525-x

29. Verma JR, Harte DSG, Shah U-K, Summers H, Thornton CA et al (2018) Investigating Flow-Sight imaging flow cytometry as a platform to assess chemically induced micronuclei using human lymphoblastoid cells in vitro.

Mutagenesis 33:283–289. https://doi.org/10.1093/mutage/gey021

30. Allemang A, Thacker R, DeMarco RA, Rodrigues MA, Pfuhler S (2021) The 3D reconstructed skin micronucleus assay using imaging flow cytometry and deep learning: proof-of-principle investigation. Mutation Res Gen Toxicol Environment Mutagenesis 865: 503314. https://doi.org/10.1016/j.mrgentox.2021.503314

31. Cheng J, Rajapakse JC (2008) Segmentation of clustered nuclei with shape markers and marking function. IEEE Transact Biomed Engineering 56:741–748. https://doi.org/10.1109/TBME.2008.2008635

32. Phansalkar N, More S, Sabale A, Joshi M (2011) Adaptive local thresholding for detection of nuclei in diversity stained cytology images. Proc Int Conf Communications and Signal Processing, In, pp 218–220. https://doi.org/10.1109/ICCSP.2011.5739305

33. Meijering E (2012) Cell segmentation: 50 years down the road. IEEE Signal Process Mag 29:140–145

34. Hernandez L, Terradas M, Martin M, Tusell L, Genesca A (2013) Highly sensitive automated method for DNA damage assessment: gamma-H2AX foci counting and cell cycle sorting. Int J Mol Sci 14:15810–15826. https://doi.org/10.3390/ijms140815810

35. Lippeveld M, Knill C, Ladlow E, Fuller A, Michaelis LJ et al (2020) Classification of human white blood cells using machine learning for stain-free imaging flow cytometry. Cytometry 97A:308–319. https://doi.org/10.1002/cyto.a.23920

36. Phillip JM, Han K-S, Chen W-C, Wirtz D, Wu P-H (2021) A robust unsupervised machine learning method to quantify the morphological heterogeneity of cells and nuclei. Nat Protocols

16:754–774. https://doi.org/10.1038/s41596-020-00432-x

37. Kumar N (2017) A dataset and a technique for generalized nuclear segmentation for computational pathology. IEEE Transact Med Imaging 36:1550–1560

38. Thibault G, Fertil B, Navarro C, Pereira S, Cau P, Levy N, Sequeira J, Mari J-L (2013) Shape and textures indexes application to cell nuclei classification. Inter J Pattern Recogn 27:1357002–1357025. https://doi.org/10.1142/S0218001413570024

39. Thibault G, Angulo J, Meyer F (2017) Advanced statistical matrices for texture chracterization application to cell nuclei classification. IEEE Transact Biomed Engineering 61:630–637. https://doi.org/10.1109/TBME.2013.2284600

40. Lee H-K, Kim C-H, Bhattacharjee S, Park H-G, Prakash D, Choi H-K (2021) A paradigm shift in nuclear chromatin interpretation: from qualitative intuitive recognition to quantitative texture analysis of breast cancer cell nuclei. Cytometry A 99A:698–706. https://doi.org/10.1002/cyto.a.24260

41. Klisch K, Pfarrer C, Schuler G, Hoffmann B, Leiser R (1999) Tripolar acytokinetic mitosis and formation of feto-maternal syncitia in the bovine placentome: different modes of the generation of multi-nuclear cells. Anat Embryol 200:229–237. https://doi.org/10.1007/s004290050275

42. Copeland M (1974) The cellular response to cytochalasin B: a critical overview. Cytologia 39:709–727

43. Riffell JL, Zimmerman C, Khong A, McHardy LM, Roberge M (2009) Effects of chemical manipulation of mitotic arrest and slippage on cancer cell survival and proliferation. Cell Cycle 8:3029–3042. https://doi.org/10.4161/cc.8.18.9623

Chapter 6

The Imaging Flow Cytometry-Based Cytokinesis-Block MicroNucleus (CBMN) Assay

Ruth C. Wilkins, Matthew Rodrigues, and Lindsay A. Beaton-Green

Abstract

The dose of ionizing radiation received by an individual can be determined using biodosimetry methods which measure biomarkers of exposure in tissue samples from that individual. These markers can be expressed in many ways, including DNA damage and repair processes. Following a mass casualty event involving radiological or nuclear material, it is important to rapidly provide this information to medical responders to assist in the medical management of potentially exposed casualties. Traditional methods of biodosimetry rely on microscope analysis, making them time-consuming and labor-intensive. To increase sample throughput following a large-scale radiological mass casualty event, several biodosimetry assays have been adapted for analysis by imaging flow cytometry. This chapter briefly reviews these methods with a focus on the most current methodology to identify and quantify micronuclei in binucleated cells within the cytokinesis-block micronucleus assay using an imaging flow cytometer.

Key words Cytokinesis-block micronucleus assay, Imaging flow cytometry, Lymphocytes, Chromosome damage, Biodosimetry

1 Introduction

Biodosimetry is a method for determining the amount of ionizing radiation an individual has been exposed to based on the measurement of biological markers. It can be used in the absence of physical dosimetry or when the physical dose estimates are in dispute. Traditionally, cytogenetic biodosimetry techniques have been used, the two most common being the dicentric chromosome assay (DCA) and the cytokinesis-block micronucleus (CBMN) assay [1]. Briefly, the DCA measures the number of dicentric chromosomes measured in human peripheral blood lymphocyte metaphase cells and converts this frequency into a dose based on an in vitro-generated dose-responsive curve. With this technique, the dicentric frequency increases with dose in a linear-quadratic manner. Similarly, the dose to an individual can be estimated using the CBMN assay, which quantifies the number of micronuclei

Natasha S. Barteneva and Ivan A. Vorobjev (eds.), *Spectral and Imaging Cytometry: Methods and Protocols*, Methods in Molecular Biology, vol. 2635, https://doi.org/10.1007/978-1-0716-3020-4_6,

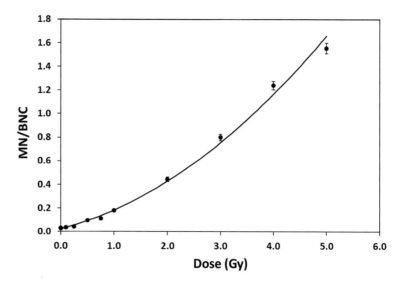

Fig. 1 Dose-responsive curve for CBMN assay generated by microscopy analysis

(MN) detected in binucleated cells (BNCs) that have been halted in cytokinesis. As with the DCA, the frequency of MN increases with increasing dose in a linear-quadratic relationship (Fig. 1). Recently γH2AX has also been identified as a marker for radiation exposure [2, 3]. This assay measures the phosphorylation of the histone H2AX which appears at the site of DNA double strand breaks. A linear relationship between increased H2AX foci and radiation dose has been observed.

Although these assays are all dose-responsive to ionizing radiation, they each have advantages and disadvantages for application to biodosimetry. Both the DCA and CBMN are standardized for biological dosimetry using manual microscopy and can generate accurate and reproducible dose estimations between 0.1 and 5 Gy [1, 4, 5]. These assays, however, suffer from a few disadvantages including unstable cytogenetic aberrations (decreasing with a half-life of about 3 years [6]), long culture times (48–72 h) prior to analysis to allow cells to cycle into mitosis where the damage is evaluated, and time-consuming and labor-intensive scoring by manual microscopy. While the γH2AX method is faster overall given that samples can be analyzed immediately without having to culture the cells, current microscope-based methods are time-consuming and prone to inter-scorer variability, and traditional flow cytometry lacks sensitivity. Additionally, one major drawback of the γH2AX assay is the rapid time kinetics – the H2AX signal increases quickly after exposure and begins to decrease after 1–2 h due to the repair of the DNA damage [2, 7, 8]. Reviews of these assays as used in biodosimetry can be found in the literature [9, 10].

Imaging flow cytometry (IFC) now makes it possible to address some of the drawbacks encountered with microscopy-based methods and to successfully adapt these assays to automated cytometry methods. Efforts have been made to adapt the DCA method to IFC [11] and a method for the measurement of dicentric chromosomes using IFC has been published previously [12]. There has also been much development in the adaptation of the CBMN assay to IFC as demonstrated by numerous publications [13–18] as well as an adapted IFC γH2AX method [19–23]. A review of these techniques as adapted to IFC for biosimetry has been published [24].

Employing these methods would significantly increase sample throughput to potentially hundreds of samples per day in a single laboratory and eventually to thousands of samples across a laboratory network. This substantial increase in throughput would improve the applicability of biosimetry as a casualty triage tool following radiological/nuclear (R/N) emergencies. This chapter will focus on the methodology developed to perform the CBMN assay using IFC.

2 IFC CBMN Method

The development of the CBMN method for flow cytometry began in the 1980s and continued into the early 2000s with some success, however, this method required lysing the cell membrane in order to measure the intensity ratio of small (MN) to large nuclei as an estimate of damage [25–27]. This removed the ability to analyse MN in BNCs as a control for proliferation as is recommended for microscope-based CBMN and made it difficult to ascertain that purported MN were legitimate [28]. In 2014, the first paper was published describing the use of IFC for the analysis of the CBMN assay [17]. With the addition of the ability to image each cell, the analysis can now be conducted on intact cells and MN frequencies can be quantified in individual BNCs. Since this first publication, much work has been completed on optimizing the assay for biosimetry and this method is now well evolved [13, 16, 18] and is starting to be applied to other fields [29, 30]. The following section describes the methods for sample preparation, staining, and image data acquisition of peripheral blood lymphocytes as well as the analysis strategy to identify and quantify BNCs and MN.

3 Materials

3.1 Lab Equipment

1. Lithium-heparin Vacutainer® tubes (10 mL).
2. Centrifuge tubes (15 mL).
3. Ventilated culture flasks (T25) (25 cm^2).

4. Eppendorf tubes (1.5 mL).

5. Imagestream (IS)X Mark II system (Luminex) with 488 nm and 642 nm lasers, (*see* **Note 1**), 40× magnification.

3.2 Reagents

1. Complete media: RPMI 1640 with 1% 2 mM L-Glutamine–Penicillin–Streptomycin solution, 10% v/v inactivated fetal bovine serum.

2. Cytochalasin B (stock solution 1.5 mg/mL; 6 µg/mL final concentration in blood culture); 2.5% phytohaemagglutinin stock solution (*see* **Note 2**).

3. Isotonic PBS (sterile): 137 mM NaCl, 2.7 mM KCl, 4.3 mM Na_2HPO_4, 1.4 mM KH_2PO_4 (*see* **Note 3**).

4. Hypotonic Solution: 75 mM KCl (*see* **Note 4**).

5. FACS Lysing solution (BD Biosciences) (*see* **Note 5**).

6. DRAQ5 (eBioscience) fluorescent probe solution (5 mM). Working DRAQ5 solution (315 µM).

4   Methods

– Unless otherwise noted, steps are conducted at room temperature (RT) which is assumed to range from 18–25 °C.

– Steps are conducted under normal lighting conditions unless otherwise noted.

4.1 Sample Preparation and Culture

1. Draw blood into lithium-heparin Vacutainers.

2. Irradiate and/or challenge samples immediately after irradiation by adding 20 µL/mL of phytohemagglutinin as required. A sham-irradiated control (0 Gy) need to be included to measure spontaneously occurring MN.

3. Dilute 2 mL of whole blood into 8 mL of complete media in a T25 flask.

4. Incubate samples for 24 h at 37 °C and 5% CO_2.

5. Add Cytochalasin-B (6 µg/mL of whole blood).

6. Incubate samples for an additional 44 h at 37 °C and 5% CO_2.

4.2 Cell Lysing and Fixation

1. After incubation, transfer cultures to 15 mL centrifuge tubes.

2. Spin at $250 \times g$ for 8 min.

3. Aspirate supernatant, leaving approximately 2 mL above the cells and without touching the buffy coat (*see* **Note 6**).

4. Resuspend the pellet thoroughly by tapping the bottom of the tube.

5. Add 10 mL of hypotonic solution and mix by inversion.

6. Incubate for 10 min.

7. Add 2 mL of 1 × FACS Lysing solution and mix by inversion.

8. Spin at 250 × g for 3 min.

9. Aspirate supernatant, leaving approximately 1 mL of supernatant above the cell pellet.

10. Resuspend the pellet thoroughly by tapping the bottom of the tube.

11. Add 10 mL 1 × FACS Lysing solution and mix by inversion.

12. Incubate for 10 min.

13. Spin at 250 × g for 3 min.

14. Aspirate supernatant.

15. Add 10 mL 1 × FACS Lysing solution and mix by inversion.

16. Incubate for 10 min.

17. Spin at 250 × g for 3 min.

18. Aspirate supernatant.

19. Add 10 mL isotonic solution.

20. Spin at 250 × g for 3 min.

21. Aspirate supernatant, leaving 0.5 mL.

22. Transfer to a 1.5 mL Eppendorf tube.

23. Spin at 250 × g for 3 min.

24. Aspirate the supernatant to about 40 μL.

25. Store at 4 °C until analysis. (*see* **Note 7**).

4.3 Staining

1. Add 5 μL of DRAQ5 working solution to all samples.

2. Mix by pipetting and incubate for 20 min in the dark.

4.4 Data Acquisition and Analysis

1. Load sample onto the ISX Mark II (*see* **Note 8**).

2. Set the ISX Mark II to 40× magnification.

3. Set the laser power and cell classifiers to optimize the collection of cells while discarding unstained debris (*see* **Note 9**).

4. Acquire samples (*see* **Note 10**).

5. Open the ISX Mark II analysis software package IDEAS (*see* **Note 11**).

6. Follow **steps 7–9** to create all masks required to identify BNCs and MN; iterate as required to fine-tune the gating strategy. A copy of the full IDEAS data analysis template is available from the authors upon request.

7. Create an overlay of the brightfield (BF) and DRAQ5 channels (channel 1 and 5 respectively in a single camera system). This permits the creation of a custom view in which the whole cell image (BF), the nuclei (DRAQ5), the BNC mask, the MN mask, and the overlay (BF/DRAQ5) can be visualized together.

8. Create a BNC mask and nuclear component masks by creating the following nested masks in IDEAS. Note that some modifications may be required depending on image quality and settings.

   (a) LevelSet (M05, DRAQ5, Middle, 3): Apply the Levelset function to the default DRAQ5 mask (M05) with the middle-intensity setting and a contour detail setting of 3. This mask creates a tight contour around the edge of the BNC.

   (b) Watershed (LevelSet (M05, DRAQ5, Middle, 3)): Apply the Watershed function to the Levelset mask in 8a. This mask creates a break in the LevelSet mask at the natural segmentation point between the two nuclei.

   (c) BNC mask: Range (Watershed (LevelSet (M05, DRAQ5, Middle, 3)), 115–5000, 0–1): Apply the Range function to the Watershed mask in 8b. Set the minimum Area value to 115 and the maximum Area value to 5000. Set the minimum Aspect Ratio value to 0 and the maximum Aspect Ratio value to 1. These settings ensure no small debris or MNis captured within the BNC mask.

   (d) Nuclear component 1: Component (1, Area, Range (Watershed (LevelSet (M05, DRAQ5, Middle, 3)), 115–5000, 0–1), Descending: Apply the Component function to the BNC mask in 8c. Choose Area as the ranking feature, set the sorting order to Descending, and set the Rank value to 1. This mask highlights the first nucleus of a BNC.

   (e) Nuclear component 2: Component (2, Area, Range (Watershed (LevelSet (M05, DRAQ5, Middle, 3)), 115–5000, 0–1), Descending: Apply the Component function to the BNC mask in 8c. Choose Area as the ranking feature, set the sorting order to Descending, and set the Rank value to 2. This mask highlights the second nucleus in a BNC.

   (f) Threshold (M05, DRAQ5, 50): Apply the Threshold function to the default DRAQ5 mask. Set the Intensity Percentage to 50. This mask will be used in the gating strategy to remove apoptotic/necrotic cells.

9. Create an MN mask by creating the following nested masks in IDEAS. Note that some modifications may be required depending on image quality and settings.

   (a) Create spot identification function 1:

      (i) Spot (M05, DRAQ5, Bright, 2, 4, 1): Apply the Spot mask to the default DRAQ5 mask (M05). Set the Spot to Cell Background Ratio to 2, the minimum radius to

1, and the maximum radius to 4 to identify candidate MN.

(ii) Range (LevelSet (M05, DRAQ5, Middle, 3), 80–5000, 0–1): Apply the Range function to the spot mask in 8a. Set the minimum area value to 80 and the maximum area value to 5000. Set the minimum Aspect Ratio value to 0 and the maximum Aspect Ratio value to 1. These settings ensure a wide range of candidate main nuclei is captured in this mask.

(iii) Dilate (Range (LevelSet (M05, DRAQ5, Middle, 3), 80–5000, 0–1), 2): Apply the Dilate function to the nuclear mask in 9aii and set the number of pixels to 2. This mask expands the nuclear mask to ensure the full outer perimeter of the nuclear portion of the image is masked. This avoids artifacts on the periphery of the nuclear image being inadvertently identified as MN.

(iv) Spot (M05, DRAQ5, Bright, 2, 4, 1) And Not Dilate (Range (LevelSet (M05, DRAQ5, Middle, 3), 80–5000, 0–1), 2): Use the "And" and "Not" Boolean algebra functions in the IDEAS mask manager to combine the masks in 9ai and 9aiii. This step subtracts the nuclear portion of the mask leaving only spots that may be candidate MN.

(v) Range (Spot (M05, Ch05, Bright, 2, 4, 1) And Not Dilate (Range (LevelSet (M05, Ch05, Middle, 3), 80–5000, 0–1), 1), 10–80, 0.4–1): Apply the Range function to the combined mask in 9aiv. Set the minimum Area value to 10 and the maximum Area value to 80. Set the minimum Aspect Ratio value to 0.4 and the maximum Aspect Ratio value to 1. This mask eliminates very small and very large spots and the values used here ensure that rounded spots adhering as closely as possible to the size requirement for scoring MN outlined in published criteria are implemented [31].

(b) Create spot identification function 2:

(i) Spot (M05, DRAQ5, Bright, 2, 4, 2): Apply the Spot mask to the default DRAQ5 mask (M05). Set the Spot to Cell Background Ratio to 2, the minimum radius to 2, and the maximum radius to 4 to identify slightly larger candidate MN than spot identification function 1.

(ii) Morphology (M05, DRAQ5): Apply the Morphology function to the default DRAQ5 mask to create a tight contour around the nuclear boundary.

(iii) Dilate (Morphology, M05, DRAQ5, 3): Apply the Dilate function to the Morphology mask in 9bii to expand the mask to completely cover the nuclear boundary. This avoids artifacts on the periphery of the nuclear image being inadvertently identified as MN.

(iv) Spot (M05, DRAQ5, Bright, 2, 4, 2) And Not Dilate (Morphology (M05, DRAQ5), 3): Use the "And" and "Not" Boolean algebra functions in the IDEAS mask manager to combine the masks in 9bi and 9biii. This step subtracts the nuclear portion of the mask leaving only spots that may be candidate MN.

(v) Range (Spot (M05, Ch05, Bright, 1, 4, 1) And Not Dilate (Morphology (M05, Ch05), 3), 6–80, 0.4–1): Apply the Range function to the combined mask in 9biv. Set the minimum area value to 6 and the maximum area value to 10. Set the minimum Aspect Ratio value to 0.4 and the maximum Aspect Ratio value to 1. This mask eliminates very small and very large spots and the values used here ensure that rounded spots adhering as closely as possible to the size requirement for scoring MN outlined in published criteria are implemented.

(c) Create the final MN mask and the MN component masks:

(i) MN mask: Range (Spot (M05, Ch05, Bright, 2, 4, 1) And Not Dilate (Range (LevelSet (M05, Ch05, Middle, 3), 80–5000, 0–1), 1), 10–80, 0.4–1) Or Range (Spot (M05, Ch05, Bright, 1, 4, 1) And Not Dilate (Morphology (M05, Ch05), 3), 6–80, 0.4–1): Use the "Or" Boolean algebra function in the IDEAS mask manager to combine the spot identification function 1 and 2 masks created in 9a and 9b. This step creates the final MN mask that ensures any spots captured by either spot identification function will be highlighted as candidate MN.

(ii) Component (1, Area, MN Mask, Descending): Apply the Component function to the MN mask in 9ci. Choose Area as the ranking feature, set the sorting order to Descending, and set the Rank value to 1. This mask highlights only the largest MN identified by the MN mask.

(iii) Component (2, Area, MN Mask, Descending): Apply the Component function to the MN mask in 9ci. Choose Area as the ranking feature, set the sorting order to Descending, and set the Rank value to

2. This mask highlights only the second-largest MN identified by the MN mask.

  (iv) Component (3, Area, MN Mask, Descending): Apply the Component function to the MN mask in 9ci. Choose Area as the ranking feature, set the sorting order to Descending, and set the Rank value to 3. This mask highlights only the third largest MN identified by the MN mask.

  (v) Component (4, Area, MN Mask, Descending): Apply the Component function to the MN mask in 9ci. Choose Area as the ranking feature, set the sorting order to Descending, and set the Rank value to 4. This mask highlights only the fourth largest MN identified by the MN mask.

10. Generate the cytoplasm mask and the skeleton mask by creating the following nested masks in IDEAS. Note that some modifications may be required depending on image quality and settings

  (a) AdaptiveErode (M01, BF, 85): Apply the AdaptiveErode function to the Default (M01) BF mask and set the Adaptive Erosion Coefficient to 85 to shrink the contour of the mask to exclude the perimeter of the BF image.

  (b) Dilate (AdaptiveErode (M01, BF, 85), 4): Apply the Dilate function to the AdaptiveErode mask in 10a and set the number of pixels to 4. This mask expands the eroded default mask in 10a and dilates it to highlight the entire cell taking its shape into account.

  (c) Dilate (AdaptiveErode (M01, BF, 85), 4) Or Component (1, Area, MN Mask, Descending) Or Component (2, Area, MN Mask, Descending) Or Component (3, Area, MN Mask, Descending) Or Component (4, Area, MN Mask, Descending): Use the "Or" Boolean algebra function in the IDEAS mask manager to create a mask that is a combination of the cytoplasm mask in 10b and the individual MN mask components in 9c. This combined mask will be used in the gating strategy to determine whether a spot is inside or outside of the cell.

  (d) Dilate (BNC, 4): Apply the Dilate function to the BNC mask created in 8c and set the number of pixels to 4.

  (e) Skeleton (Dilate (BNC, 4), DRAQ5, Thin): Apply the Skeleton function to the dilated BNC mask in 10d and select the 'Thin' radio button. This creates a line mask along the center of the major axis of the BNC.

(f) Dilate (Skeleton (Dilate (BNC, 4), DRAQ5, Thin), 1): Apply the Dilate function to the Skeleton mask in 10e. This increases the thickness of the line mask.

(g) Dilate (Skeleton (Dilate (BNC, 4), DRAQ5, Thin), 1) Or Component (1, Area, MN Mask, Descending) Or Component (2, Area, MN Mask, Descending) Or Component (3, Area, MN Mask, Descending) Or Component (4, Area, MN Mask, Descending): Use the "Or" Boolean algebra function in the IDEAS mask manager to create a mask that is a combination of the skeleton line mask in 10f and the individual MN mask components in 9c. This combined mask will be used in the gating strategy to remove any artifacts between the two nuclei of the BNC that are inadvertently highlighted by the MN mask.

11. Follow this step to create all features and regions in the gating strategy used to identify BNCs. Additional modifications may be required depending on image quality and settings (*see* **Note 12**). One gating strategy for identifying BNCs and MN has been previously well described [15]. Modifications have been made in recent years and the most current gating strategy is described here. This is a sequential gating strategy, where each step uses the prior step as the new parent population unless otherwise noted.

(a) Plot BF Area versus BF Aspect Ratio using all events as the parent population. Gate on single cells by setting the BF Area region to range from 50 to 600 and the BF Aspect Ratio region to range from 0.6 to 1.

(b) Plot BF Gradient RMS versus DRAQ5 Gradient RMS. Gate on focused cells by setting the BF Gradient RMS region to range from 45 to 75 and the DRAQ5 Gradient RMS region to range from 14 to 26.

(c) Plot a histogram of DRAQ5 Raw Max Pixel. Gate on cells with sufficient DRAQ5 staining by creating a linear region that ranges from 1000 to 4095 (*see* **Note 9**).

(d) Create the Area Threshold 50% feature by applying the Area Threshold 50% mask to the Area feature. Plot BF Contrast versus Area Threshold 50%. Gate on non-apoptotic cells by setting the BF Contrast region to range from 0 to 30 and the Area Threshold 50% region to range from 50 to 300.

(e) Apply the BNC mask to the Spot Count feature to create the BNC Spot Count feature. Plot a histogram of the BNC Spot count feature and draw a linear region from 1.5 to 2.5 to select cells that have two nuclei.

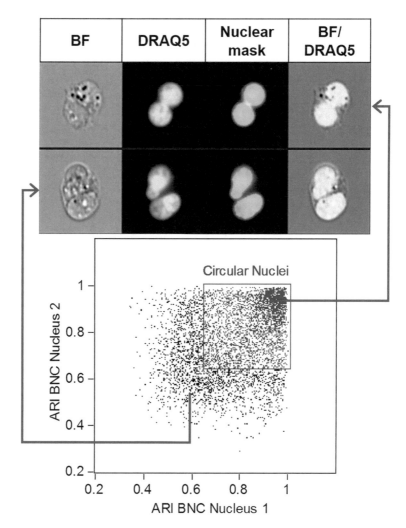

Fig. 2 Bivariate plot of the Aspect Ratio Intensity feature computed on both nuclei to select for BNCs having two highly circular nuclei

(f) Apply the first nuclear component mask (8d) to the Aspect Ratio Intensity (ARI) feature to create the ARI Nuclear Component 1 feature. Apply the second nuclear component mask (8e) to the ARI feature to create the ARI Nuclear Component 2. Plot these two features on a bivariate plot to gate on BNCs in which both nuclei are highly circular by round by setting both region boundaries to range from 0.65 to 1 (Fig. 2).

(g) Create a combined Area Ratio feature by dividing the Area of the second nuclear component by the Area of the first nuclear component. Create a combined Intensity Ratio feature by creating two new features and computing their quotient:

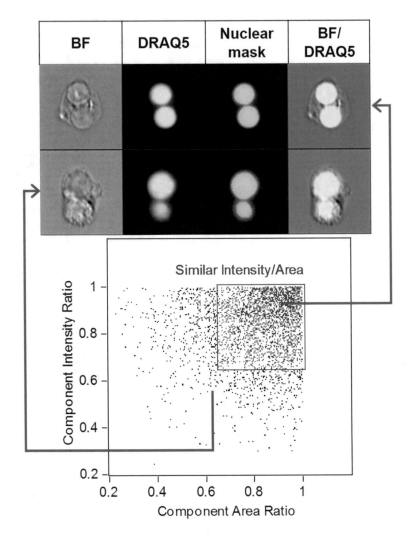

Fig. 3 Bivariate plot of the Component Area Ratio feature versus the Component Intensity Ratio to simultaneously select for BNCs having two nuclei of similar area and similar intensity

    (i) Apply the BNC mask to the Spot Intensity Minimum (SIMin) feature

    (ii) Apply the BNC mask to the Spot Intensity Maximum (SIMax) feature

    (iii) Create the Intensity Ratio feature by dividing SIMin (11gi) by SIMax (11gii)

(h) Plot the Area Ratio feature versus the Intensity Ratio feature to gate on BNCs in which both nuclei have similar areas and intensities by setting both region boundaries to range from 0.65 to 1 (Fig. 3).

(i) Apply the BNC mask to both the Aspect Ratio and Shape Ratio features. Plot these two features on a bivariate plot and gate on BNCs with two distinct, non-overlapping nuclei by setting both region boundaries to range from 0.55 to 3.

(j) To gate on the final BNC population, create two new combined features:

(i) Apply the BNC mask to the Area feature and apply the DRAQ5 default mask to the Area feature. Use the mathematical operators in the combined feature window to create a combined feature by dividing the Area BNC feature by the Area DRAQ5 feature.

(ii) Apply the first nuclear component mask (8d) to the Circularity feature. Apply the second nuclear component mask (8e) to the Circularity feature. Use the mathematical operators in the combined feature window to create an average BNC circularity feature as follows: (Circularity nuclear component 1 + Circularity nuclear component 2)/2.

(k) Plot these two combined features against one another on a bivariate plot and gate on BNCs that have two highly circular nuclei and minimal background staining by setting the Area BNC/Area DRAQ5 region to range from 0.275 to 0.6 and the Circularity BNC region to range from 7 to 25.

12. Follow this step to create all features and regions in the gating strategy used to identify MN within BNCs.

(a) Apply the MN mask to the Spot Count feature to create the MN Spot Count feature. Plot a histogram of the MN Spot count feature and draw a linear region from 0.5 to 4.5 to select all cells that have potential MN (Fig. 4).

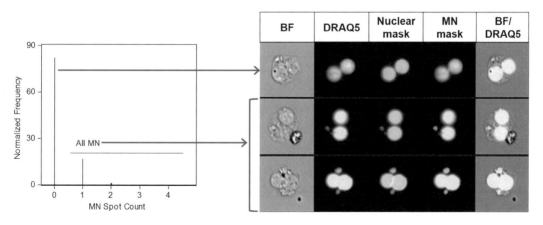

Fig. 4 Histogram of the Spot Count feature applied to the MN mask to identify all BNCs with any number of candidate MN

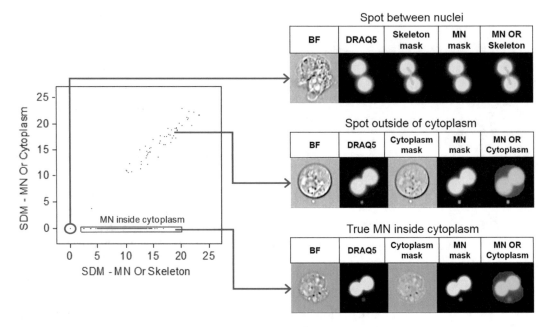

Fig. 5 Bivariate plot of the Spot Distance Minimum (SDM) feature applied to both the cytoplasm mask and the skeleton mask. The skeleton mask permits removal of false positive artifacts residing between the two nuclei (top image) while the cytoplasm mask allows elimination of BNCs with MN or other spots outside of the cytoplasm (middle image)

(b) Create two new features:

(i) Apply the Adaptive Erode or MN Components mask from 10c to the Spot Distance Minimum feature.

(ii) Apply the Skeleton or MN Components mask from 10*g* to the Spot Distance Minimum feature.

(c) Plot the feature in (12bi) on the *x*-axis and plot the feature in (12bii) on the *y*-axis of a bivariate plot. Gate on BNCs that have MN inside the cytoplasm while removing artifacts that may be masked between the nuclei by setting the *x*-axis feature region to range from 0.2 to 10 and the *y*-axis feature to range from −0.05 to 0.05 (Fig. 5).

(d) Apply the MN mask to the Spot Count feature to create the MN Spot Count feature. Plot a histogram of the MN Spot count feature and draw four linear regions:

(i) From 0.8 to 1.2 to select cells that have one MN

(ii) From 1.8 to 2.2 to select cells that have two MN

(iii) From 2.8 to 3.2 to select cells that have three MN

(iv) From 3.8 to 4.2 to select cells that have four MN

(e) Identify cells with one or true MN: Apply the 1 MN Component mask (9cii) to the Aspect Ratio feature to

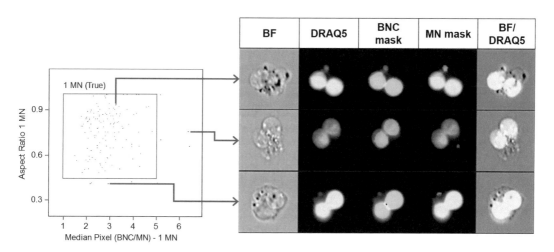

Fig. 6 Bivariate plot of the Aspect Ratio feature applied to the MN mask versus the ratio of the Medial Pixel features applied to both the BNC and MN masks to eliminate elongated spots and dim artifacts or MN

create the Aspect Ratio 1 MN feature. Create two new features and compute their quotient:

(i) Apply the BNC mask to the Median Pixel feature using the DRAQ5 image

(ii) Apply the 1 MN Component mask (9cii) to the Median Pixel feature using the DRAQ5 image

(iii) Create a combined Median Pixel Ratio 1 MN feature by dividing 12di by 12dii

(f) Plot the Median Pixel Ratio 1 MN feature on the *x*-axis versus the Aspect Ratio 1 MN feature on the *y*-axis of a bivariate plot. Gate on MN that are circular and of similar intensity to the main nuclei of the BNCs setting the *x*-axis feature region to range from 1 to 5 and the *y*-axis feature to range from 0.45 to 1 (Fig. 6).

(g) Identify cells with one or two MN: Apply the 2 MN Component mask (9ciii) to the Aspect Ratio feature to create the Aspect Ratio 2 MN feature. Create the following new features:

(i) Use the mathematical operators in the combined feature window to create an Average Aspect Ratio 1 + 2 MN feature as follows: (Aspect Ratio 1 MN + Aspect Ratio 2 MN)/2.

(ii) Use the mathematical operators in the combined feature window to create an Average Median Pixel MN 1 + 2 feature as follows: (Median Pixel 1 MN + Median Pixel 2 MN)/2.

       (iii) Create a new combined Median Pixel Ratio 1 + 2 MN feature by dividing the Median Pixel BNC feature (12di) by the Average Median Pixel MN 1 + 2 (12eii).

(h) Plot the Median Pixel Ratio 1 + 2 MN (12eiii) feature on the x-axis versus the Average Aspect Ratio 1 + 2 MN feature (12ei) on the y-axis of a bivariate plot. Gate on cells that have two MN that are circular and of similar intensity to the main nuclei of the BNCs by setting the x-axis feature region to range from 1 to 5 and the y-axis feature to range from 0.45 to 1.

(i) Identify cells with three true MN: Apply the 3 MN Component mask (9civ) to the Aspect Ratio feature to create the Aspect Ratio 3 MN feature. Create the following new features:

     (i) Use the mathematical operators in the combined feature window to create an Average Aspect Ratio 1 + 2 + 3 MN feature as follows: (Aspect Ratio 1 MN + Aspect Ratio 2 MN + Aspect Ratio 3 MN)/3.

     (ii) Use the mathematical operators in the combined feature window to create an Average Median Pixel MN 1 + 2 + 3 feature as follows: (Median Pixel 1 MN + Median Pixel 2 MN + Median Pixel 3 MN)/3.

     (iii) Create a new combined Median Pixel Ratio 1 + 2 + 3 MN feature by dividing the Median Pixel BNC feature by the Average Median Pixel MN 1 + 2 + 3.

(j) Plot the Median Pixel Ratio 1 + 2 + 3 MN (12fiii) feature on the x-axis versus the Average Aspect Ratio 1 + 2 + 3 MN feature (12fi) on the y-axis of a bivariate plot. Gate on cells that have three MN that are circular and of similar intensity to the main nuclei of the BNCs by setting the x-axis feature region to range from 1 to 5 and the y-axis feature to range from 0.45 to 1.

(k) Identify cells with four true MN: Apply the 4 MN Component mask (9civ) to the Aspect Ratio feature to create the Aspect Ratio 4 MN feature. Create the following new features:

     (i) Use the mathematical operators in the combined feature window to create an Average Aspect Ratio 1 + 2 + 3 + 4 MN feature as follows: (Aspect Ratio 1 MN + Aspect Ratio 2 MN + Aspect Ratio 3 MN + Aspect Ratio 4 MN)/4.

     (ii) Use the mathematical operators in the combined feature window to create an Average Median Pixel MN 1 + 2 + 3 + 4 feature as follows: (Median Pixel

1 MN + Median Pixel 2 MN + Median Pixel 3 MN + Aspect Ratio 4 MN)/4.

(iii) Create a new combined Median Pixel Ratio 1 + 2 + 3 + 4 MN feature by dividing the Median Pixel BNC feature by the Average Median Pixel MN 1 + 2 + 3 + 4.

(l) Plot the Median Pixel Ratio 1 + 2 + 3 + 4 MN (12giii) feature on the *x*-axis versus the Average Aspect Ratio 1 + 2 + 3 + 4 MN feature (12gi) on the *y*-axis of a bivariate plot. Gate on cells that have four MN that are circular and of similar intensity to the main nuclei of the BNCs by setting the *x*-axis feature region to range from 1 to 5 and the *y*-axis feature to range from 0.45 to 1.

13. Create a statistics report by selecting the Reports tab and selecting Define Statistics Report. Create the following columns:

(a) BNC count: Select "Count" as the statistic and select BNCs (11i) as the population

(b) 1 MN count: Select "Count" as the statistic and select 1 MN (12d) as the population

(c) 2 MN count: Select "Count" as the statistic and select 2 MN (12e) as the population

(d) 3 MN count: Select "Count" as the statistic and select 3 MN (12f) as the population

(e) 4 MN count: Select "Count" as the statistic and select 4 MN (12g) as the population

5   Notes

1. If a 642 nm laser is not available, the 488 nm can be used with DRAQ5. Alternatively, the method can be adapted for use with a different laser and an alternative DNA stain such as Hoechst 33342 which is excited by the 405 nm laser.

2. Make a working solution fresh on the first day of the experiment. Let warm in an incubator at 37 °C and 5% CO_2 until needed.

3. pH 7.4 in ddH_2O and filter sterile. Can be stored at RT for up to a year.

4. pH 7.4 in ddH_2O and filter (0.2 μm) and store at RT for up to several months.

5. Dilute 1:10 in deionized water on the day of the experiment.

6. The buffy coat will form a white layer on top of the red cells.

7. If samples are to be stored for more than 2 days, it is best to leave more isotonic solution and spin down and aspirate to 40 μL on the day of analysis.

8. Samples can also be run on a 96-well plate using the autosampler.

9. The following acquisition settings were used in our laboratory: brightfield LED on, a 642 nm laser set to an appropriate laser power such that the main peak on a DRAQ5 Raw Max Pixel histogram fell between 1000 and 3000. Images were collected on a single camera system from channel 1 (brightfield signal, LED) and channel 5 (DRAQ5). Data were collected using the INSPIRE software (version 200.1.388.0) with only the Area feature applied. Events with an area less than 100 pixels (25 μm^2) were gated out to minimize the collection of small debris. Note that these settings will vary by machine, staining intensity, and cell type, and will need to be optimized for different experiments. If the population of acquired cells falls outside the Raw Max Pixel region between 1000 and 3000, the gating will need to be adjusted accordingly to ensure cells of interest are included.

10. The total number of objects to acquire such that sufficient numbers of BNCs will be imaged will vary based on factors such as sample concentration. It is recommended that smaller data files are collected initially from the "All" population to determine the approximate location of BNCs on the Area versus Aspect Ratio and Gradient RMS bivariate plots. Once the acquisition settings are optimized, then larger data files can be collected as a large number of BNCs will be captured in the data file while simultaneously minimizing the collection of debris and events that are not of interest in the CBMN assay and keeping the file size small. Typically, in our laboratory, once classifiers were set, 100,000 cells were acquired.

11. Samples were analyzed uncompensated after testing whether compensation improved the results. Since this is a one-color assay using only the DRAQ5 nuclear stain, spectral overlap into the nuclear image was not a factor and therefore compensation was unnecessary. This should be tested based on the conditions used and should additional fluorochromes be used in the assay, proper single color controls to generate a compensation matrix must be collected.

12. Gate boundaries shown in brackets after each feature are as a guide only. Note that to optimize the gating strategy in the IDEAS software, it is important to ensure that the region boundaries are not gating out scorable events. Using the population tagging feature, the user can define populations of objects of interest (i.e., legitimate BNCs) as well as populations

that the user would like to gate out (i.e., BNCs that do not fit the published scoring criteria). These populations can be displayed on any bivariate plot, permitting the region boundaries to be optimized [15].

Disclaimer

The development of this method was initiated at Health Canada. A collaboration was then set up with Luminex Corporation to further develop the masking protocol and analysis template. Health Canada was internally funded for this work. Matthew A. Rodrigues is employed by Luminex Corporation, the maker of the Amnis® brand ImageStream® imaging flow cytometer used in this work.

References

1. IAEA (2011) Cytogenetic dosimetry: applications in preparedness for and response to radiation emergencies. EPR-Biodosimetry. International Atomic Energy Agency, Vienna, p 246

2. Rothkamm K, Horn S (2009) Gamma-H2AX as protein biomarker for radiation exposure. Ann Ist Super Sanita 45:265–271

3. Horn S, Barnard S, Rothkamm K (2011) Gamma-H2AX-based dose estimation for whole and partial body radiation exposure. PLoS One 6:e25113

4. International Organization for Standardization (2014) Radiation protection - performance criteria for laboratories using the cytokinesis-blocked micronucleus assay in blood lymphocytes for biological dosimetry. ISO, Geneva

5. International Organization for Standardization (2014) Radiation protection – perfomance criteria for service laboratories performing biological dosimetry by cytogenetics. ISO, Geneva

6. Lloyd DC, Purrott RJ, Reeder EJ (1980) The incidence of unstable chromosome aberrations in peripheral blood lymphocytes from unirradiated and occupationally exposed people. Mutat Res 72:523–532

7. Vandersickel V, Beukes P, Van BB, Depuydt J, Vral A, Slabbert J (2014) Induction and disappearance of gammaH2AX foci and formation of micronuclei after exposure of human lymphocytes to (6)(0)Co gamma-rays and p(66)+Be(40) neutrons. Int J Radiat Biol 90:149–158

8. Rogakou EP, Boon C, Redon C, Bonner WM (1999) Megabase chromatin domains involved in DNA double-strand breaks in vivo. J Cell Biol 146:905–916

9. Ainsbury EA, Barquinero JF (2009) Biodosimetric tools for a fast triage of people accidentally exposed to ionising radiation. Statistical and computational aspects. Ann Ist Super Sanita 45:307–312

10. Bender MA, Awa AA, Brooks AL, Evans HJ, Groer PG, Littlefield LG, Pereira C, Preston RJ, Wachholz BW (1988) Current status of cytogenetic procedures to detect and quantify previous exposures to radiation. Mutat Res 196:103–159

11. Beaton-Green LA, Rodrigues MA, Lachapelle S, Wilkins RC (2017) Foundations of identifying individual chromosomes by imaging flow cytometry with applications in radiation biodosimetry. Methods 112:18–24

12. Beaton-Green LA, Wilkins RC (2016) Quantitation of chromosome damage by imaging flow cytometry. In: Bartenva NS, Vorobjev IA (eds) Imaging flow cytometry, Methods in molecular biology, vol 1389. Humana-Press, pp 97–110. https://doi.org/10.1007/978-1-4939-3302-0_6

13. Rodrigues MA, Beaton-Green LA, Wilkins RC (2016) Validation of the cytokinesis-block micronucleus assay using imaging flow cytometry for high throughput radiation biodosimetry. Health Phys 110:29–36. https://doi.org/10.1097/HP.0000000000000371

14. Rodrigues MA, Probst CE, Beaton-Green LA, Wilkins RC (2016) The effect of an optimized imaging flow cytometry analysis template on sample throughput in the reduced culture cytokinesis-block micronucleus assay. Radiat Prot Dosim 172:223–229. https://doi.org/10.1093/rpd/ncw160

15. Rodrigues MA, Probst CE, Beaton-Green LA, Wilkins RC (2016) Optimized automated data analysis for the cytokinesis-block micronucleus assay using imaging flow cytometry for high throughput radiation biodosimetry. Cytometry A 89:653–662. https://doi.org/10.1002/cyto.a.22887

16. Wang Q, Rodrigues MA, Repin M, Pampou S, Beaton-Green LA, Perrier J, Garty G, Brenner DJ, Turner HC, Wilkins RC (2019) Automated triage radiation biodosimetry: integrating imaging flow cytometry with high-throughput robotics to perform the cytokinesis-block micronucleus assay. Radiat Res 191:342–351

17. Rodrigues MA, Beaton-Green LA, Kutzner BC, Wilkins RC (2014) Automated analysis of the cytokinesis-block micronucleus assay for radiation biodosimetry using imaging flow cytometry. Radiat Environ Biophys 53:273–282. https://doi.org/10.1007/s00411-014-0525-x

18. Rodrigues MA, Beaton-Green LA, Kutzner BC, Wilkins RC (2014) Multi-parameter dose estimations in radiation biodosimetry using the automated cytokinesis-block micronucleus assay with imaging flow cytometry. Cytometry A 85:883–893. https://doi.org/10.1002/cyto.a.22511

19. Parris CN, Adam ZS, Al-Ali H, Bourton EC, Plowman C, Plowman PN (2015) Enhanced gamma-H2AX DNA damage foci detection using multimagnification and extended depth of field in imaging flow cytometry. Cytometry A 87:717–723

20. Bourton EC, Plowman PN, Zahir SA, Senguloglu GU, Serrai H, Bottley G, Parris CN (2012) Multispectral imaging flow cytometry reveals distinct frequencies of gamma-H2AX foci induction in DNA double strand break repair defective human cell lines. Cytometry A 81:130–137

21. Garty G, Chen Y, Salerno A, Turner H, Zhang J, Lyulko O, Bertucci A, Xu Y, Wang H, Simaan N, Randers-Pehrson G, Yao YL, Amundson SA, Brenner DJ (2010) The RABIT: a rapid automated biodosimetry tool for radiological triage. Health Phys 98:209–217

22. Lee Y, Wang Q, Shuryak I, Brenner DJ, Turner HC (2019) Development of a high-throughput γ-H2AX assay based on imaging flow cytometry. Radiat Oncol 14:1–10.

https://doi.org/10.1186/s13014-019-1344-7

23. Durdik M, Kosik P, Gursky J, Vokalova L, Markova E, Belyaev I (2015) Imaging flow cytometry as a sensitive tool to detect low-dose-induced DNA damage by analyzing 53BP1 and γH2AX foci in human lymphocytes. Cytometry A 87:1070–1078. https://doi.org/10.1002/cyto.a.22731

24. Wilkins RC, Rodrigues MA, Beaton-Green LA (2017) The application of imaging flow cytometry to high-throughput biodosimetry. Genome Integr 8:1–7. https://doi.org/10.4103/2041-9414.198912

25. Nusse M, Marx K (1997) Flow cytometric analysis of micronuclei in cell cultures and human lymphocytes: advantages and disadvantages. Mutat Res 392:109–115

26. Avlasevich SL, Bryce SM, Cairns SE, Dertinger SD (2006) In vitro micronucleus scoring by flow cytometry: differential staining of micronuclei versus apoptotic and necrotic chromatin enhances assay reliability. Environ Mol Mutagen 47:56–66

27. Dertinger SD, Miller RK, Brewer K, Smudzin T, Torous DK, Roberts DJ, Avlasevich SL, Bryce SM, Sugunan S, Chen Y (2007) Automated human blood micronucleated reticulocyte measurements for rapid assessment of chromosomal damage. Mutat Res 626:111–119

28. Fenech M, Morley AA (1985) Measurement of micronuclei in lymphocytes. Mutat Res 147:29–36

29. Rodrigues MA (2019) An automated method to perform the in vitro micronucleus assay using multispectral imaging flow cytometry. J Vis Exp 2019:1–13. https://doi.org/10.3791/59324

30. Verma JR, Harte DSG, Shah UK, Summers H, Thornton CA, Doak SH, Jenkins GJS, Rees P, Wills JW, Johnson GE (2018) Investigating FlowSight® imaging flow cytometry as a platform to assess chemically induced micronuclei using human lymphoblastoid cells in vitro. Mutagenesis 33:283–289. https://doi.org/10.1093/mutage/gey021

31. Fenech M, Chang WP, Kirsch-Volders M, Holland N, Bonassi S, Zeiger E (2003) HUMN project: detailed description of the scoring criteria for the cytokinesis-block micronucleus assay using isolated human lymphocyte cultures. Mutat Res 534:65–75

Chapter 7

High-Throughput γ-H2AX Assay Using Imaging Flow Cytometry

Younghyun Lee, Qi Wang, Ki Moon Seong, and Helen C. Turner

Abstract

The γ-H2AX assay is a sensitive and reliable method to evaluate radiation-induced DNA double-strand breaks. The conventional γ-H2AX assay detects individual nuclear foci manually, but is labor-intensive and time-consuming, and hence unsuitable for high-throughput screening in cases of large-scale radiation accidents. We have developed a high-throughput γ-H2AX assay using imaging flow cytometry. This method comprises (1) sample preparation from small volumes of blood in the Matrix™ 96-tube format, (2) automated image acquisition of cells stained with immunofluorescence-labeled γ-H2AX using ImageStream®^{X}, and (3) quantification of γ-H2AX levels and batch processing using the Image Data Exploration and Analysis Software (IDEAS®). This enables the rapid analysis of γ-H2AX levels in several thousand of cells from a small volume of blood with accurate and reliable quantitative measurements for γ-H2AX foci and mean fluorescence levels. This high-throughput γ-H2AX assay could be a useful tool not only for radiation biodosimetry in mass casualty events, but also for large-scale molecular epidemiological studies and individualized radiotherapy.

Key words Imaging flow cytometry, γ-H2AX, Double-strand break, High-throughput platform, Radiation

1 Introduction

Radiation-induced DNA double-strand breaks (DSBs) are serious DNA lesions that cause genomic instability, mutations, chromosomal aberrations, and eventually lead to cancer [1]. As an early cellular response to DSBs, serine 139 of the histone protein H2AX is rapidly phosphorylated at nascent DSB sites, forming γ-H2AX [2]. The immunofluorescence-based γ-H2AX assay allows visualization and quantification of discrete γ-H2AX foci, and has emerged as a reliable and sensitive method to detect radiation-induced DSBs [3]. The kinetics of γ-H2AX formation may reflect the DSB repair capacity and individual radiosensitivity; therefore, the γ-H2AX assay may be potentially useful for risk assessment and have clinical implications [4].

Natasha S. Barteneva and Ivan A. Vorobjev (eds.), *Spectral and Imaging Cytometry: Methods and Protocols*,
Methods in Molecular Biology, vol. 2635, https://doi.org/10.1007/978-1-0716-3020-4_7,
© Springer Science+Business Media, LLC, part of Springer Nature 2023

There is a need to quickly estimate received doses of potentially exposed victims in a large-scale radiological incident. The γ-H2AX assay is considered useful for early biodosimetry from hours to ~ 3 days post-exposure [5–8]. Since the γ-H2AX assay has a relatively rapid processing time of a few hours post blood sampling, it is a good candidate for high-throughput biodosimetry or DSB monitoring [5, 8]. Indeed, γ-H2AX has the potential to be used as a biomarker for exposure to ionizing radiation as low as 1 mGy [9]. Previously, the most common approach is to manually count γ-H2AX foci either directly using a microscope or using images previously captured. The method is laborious, time-consuming, and subject to human error [10]. Therefore, adapting the conventional method for high-throughput screening in cases of mass casualty events or its use in clinical settings is difficult.

Various efforts have been made to develop flow cytometry-based γ-H2AX assays that eliminate the labor-intensive and time-consuming steps of the manual method [10–12], but a limitation has been that they quantify only overall γ-H2AX fluorescence intensity, but not the foci number. Imaging flow cytometry is a tool that combines flow cytometry and conventional microscopy, enabling high-throughput characterization of cells at a microscopic scale [13–15]. It allows fast and accurate quantification of both the γ-H2AX foci number and the fluorescence levels in thousands of cells. With the batch processing option, multiple sample data files can be handled at one time using the same criteria for γ-H2AX quantification.

Recently, we developed a rapid, high-throughput γ-H2AX assay using imaging flow cytometry [16]. This chapter describes the complete protocol, from cell culture to data analysis. This protocol would be useful for evaluating radiation-induced risk in large-scale radiological accidents. Additionally, this approach could be applied in other clinical and research fields, such as in investigations of individual radiosensitivity and individual adverse effects of radiotherapy.

2   Materials

2.1   Lab Equipment

1. Lithium-heparinized vacutainer tubes (BD Biosciences, NJ, USA) (*see* **Note 1**).

2. 1.4 mL 2D Matrix™ microtubes and rack (Thermo Fisher Scientific, MA, USA).

3. Cell culture incubator (37°, 5% CO_2).

4. Eppendorf microtubes (1.5 mL).

5. 1.2 mL multichannel electronic pipette (*see* **Note 2**).

6. ImageStream®^{X} (Amnis-Luminex Corporation, TX, USA) equipped with a 40× objective and 488 nm laser.

7. INSPIRE® data acquisition software (Amnis-Luminex Corporation).

8. IDEAS® software (Amnis-Luminex Corporation).

9. Tabletop centrifuge.

2.2 Reagents

1. Complete medium: RPMI 1640 (Gibco®, Thermo Fisher Scientific) supplemented with 15% FBS and 2% Penicillin/Streptomycin (Invitrogen™, Thermo Fisher Scientific).

2. Lyse/Fix solution (5×, BD Phosflow™, BD Biosciences, San Jose, CA, USA): diluted 1× (v/v) with distilled water.

3. 1× PBS (Gibco®, Thermo Fisher Scientific).

4. Methanol: diluted to 50% (v/v) with PBS.

5. Triton X-100 (Sigma-Aldrich, St. Louis, MO, USA): diluted to 0.1% (v/v) with PBS.

6. Bovine serum albumin (BSA, Sigma-Aldrich): 1% (w/v) solution in PBS.

7. Alexa Fluor 488 mouse anti-H2AX (pS139) antibody (clone N1-431, BD Pharmingen™, BD Biosciences): diluted 1:500 (v/v, stock solution) with 1% BSA.

8. DRAQ5™ (Thermo Scientific™, Thermo Fisher Scientific) 5 mM: diluted to 50 μM in 1× PBS.

---

3 Methods

This method comprises (1) sample preparations of small volumes of blood in a Matrix™ 96-tube format, (2) automated image acquisition of cells stained with immunofluorescence-labeled γ-H2AX using ImageStream®^{X} and INSPIRE® software, and (3) quantification of γ-H2AX levels and batch processing using the IDEAS® Software (Fig. 1). Figures in this chapter are produced using blood cells from healthy donors with informed consent [16].

3.1 Sample Preparation and Culture

1. Collect blood by venipuncture into a heparinized vacutainer tube (*see* **Notes 3–4**).

2. Transfer blood aliquots (100 μL) to matrix tubes containing 900 μL complete medium (*see* **Note 5**).

3. Incubate the matrix tubes in a 37 °C, 5% CO_2 incubator as required (*see* **Note 6**).

3.2 Lysis and Fixation

1. Centrifuge tubes at 250g for 5 min at room temperature (RT) (*see* **Note 7**).

2. Remove 950 μL of the supernatant (leaving 50 μL), add 950 μL 1× lyse/fix buffer, and mix. Incubate at 37 °C for 10 min, followed by centrifugation at 250g for 5 min at RT.

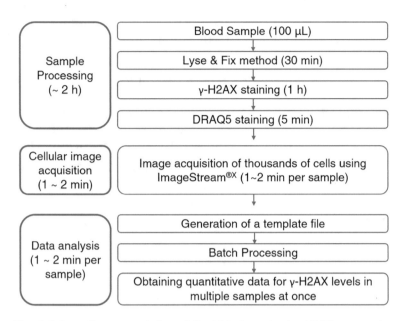

Fig. 1 Schematic representation of the high-throughput γ-H2AX assay using imaging flow cytometry. This assay consists of sample processing, image acquisition, and data analysis steps. Imaging flow cytometry enables the rapid analysis of γ-H2AX levels in thousands of cells. In addition, the lysis & fixation method and batch processing increases the efficiency of γ-H2AX assay

3. Wash 1: discard 950 μL of the supernatant (leaving 50 μL), add 950 μL PBS, resuspend the pellet, and centrifuge the tubes at 250*g* for 5 min at RT.

4. Wash 2: discard 950 μL of the supernatant (leaving 50 μL), add 950 μL PBS, resuspend the pellet, and centrifuge tubes at 250*g* for 5 min at RT.

5. Discard 950 μL of the supernatant (leaving 50 μL).

6. Add 950 μL of 50% (v/v) cold methanol and store at 4 °C.

3.3 Cell Staining

1. Centrifuge the tubes prepared in the previous section at 250*g* for 5 min at RT.

2. Wash the pellet by discarding 950 μL of the supernatant (leaving 50 μL), adding 950 μL PBS, resuspending the pellet, and centrifuging the tube at 250*g* for 5 min at RT.

3. Discard 950 μL of the supernatant (leaving 50 μL).

4. For permeabilization, add 950 μL 0.1% Triton X-100, mix, and incubate at RT for 10 min.

5. Centrifuge the tubes at 250*g* for 5 min at RT.

6. Wash the pellet by discarding 950 μL of the supernatant (leaving 50 μL), adding 950 μL PBS, resuspending the pellet, and centrifuging the tubes at 250*g* for 5 min at RT.

7. Discard 950 μL of the supernatant (leaving 50 μL).

8. Add 50 μL γ-H2AX antibody stock, mix, and incubate at RT for 1 h (*see* **Note 8**).

9. Wash 1: Add 900 μL PBS, resuspend the pellet, and centrifuge at 250*g* for 5 min at RT.

10. Wash 2: Remove 950 μL of the supernatant (leaving 50 μL), add 950 μL PBS, resuspend the pellet, and centrifuge at 250*g* for 5 min at RT.

11. Remove 950 μL of the supernatant (leaving 50 μL), resuspend the pellet, and transfer the sample into a 1.5 mL Eppendorf tube.

12. Take 50 μL of the cell suspension and add 5 μL of 50 μM DRAQ5 (final concentration: 5 μM).

13. Mix and incubate at RT for a minimum of 5 min.

3.4 Data Acquisition and Analysis

1. Start the ImageStream®^{X} and run the INSPIRE® software. Set the laser power based on the raw max pixel feature for the fluorescence channel and cell classifiers, to optimize collection of single cells (Fig. 2; *see* **Note 9**).

2. Run a sample prepared in the previous section, and acquire images of the experimental samples (*see* **Notes 10–11**).

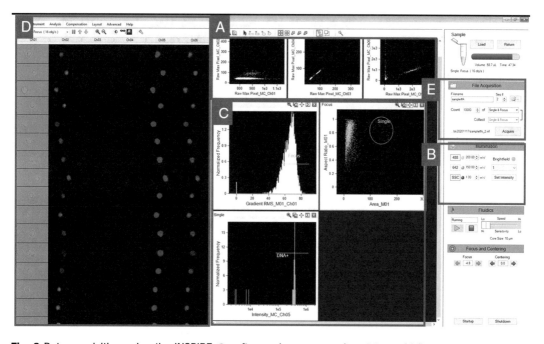

Fig. 2 Data acquisition using the INSPIRE ® software. Laser power is set to avoid fluorescence saturation based on Max Raw pixel values (**a**). We use 488 nm laser at 200 mW (**b**). Optimal gates to collect well-focused and single cells are prepared (**c**) by checking the real-time image gallery (**d**). Around 10,000 single cell events are collected (**e**)

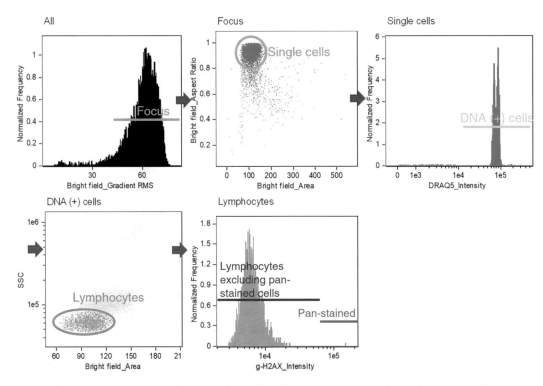

Fig. 3 Hierarchical gating strategy for selecting cells of interest. Cells are gated using Gradient RMS of the bright field image. Single cells are gated using the area and aspect ratio of bright field images. DNA-positive cells are gated using the intensity value of DRAQ5. Lymphocytes are gated using the area of bright field and side scatter, and cells with pan-nuclear γ-H2AX staining are gated out

3. Open the ImageStream analysis software package IDEAS®.

4. Follow **steps 5–9** to identify non-apoptotic single cells (Fig. 3).

5. Set a region marker to identify focused cells based on the gradient root mean square (RMS) of the Bright field (BF) image.

6. Identify single cells based on the area and aspect ratio of the BF image and gate accordingly.

7. Set a region marker to identify DNA-positive cells based on the intensity of DRAQ5 fluorescence.

8. Identify lymphocytes based on the area of BF images and the intensity of Side Scatter (SSC) and gate accordingly (*see* **Note 12**).

9. Set a region marker to identify pan-nuclear γ-H2AX-stained cells based on the fluorescence intensity of γ-H2AX. The cell population excluding pan-stained cells is used for γ-H2AX analysis (*see* **Note 13**).

10. Follow **steps 11–15** to generate a mask to detect and analyze the γ-H2AX fluorescence signal (Fig. 4).

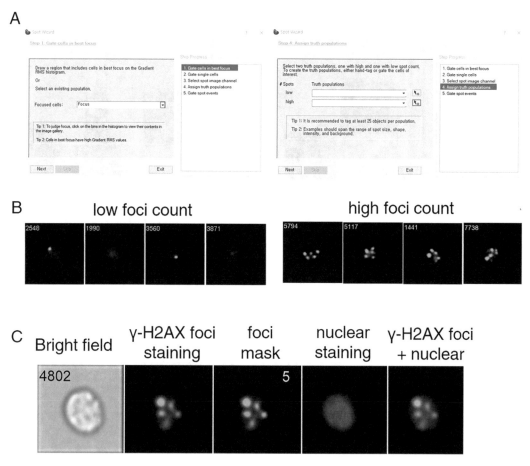

Fig. 4 Generation of spot mask using Spot Wizard in the IDEAS software. The cell population of interest is selected using Spot Wizard. To generate mask for spot counting, true populations with low and high foci should be selected (**a**). Representative images of cells with low and high foci are displayed (**b**). Based on images of the selected true population, the spot mask is automatically generated. Images (**c**) show nuclear and γ-H2AX image with foci mask. The foci are identified by the mask "Range(Peak(Spot(M02, Channel 2, Bright, 4.5, 1), Channel 2, Bright,0), 0-200,0-1)". The number at the top right corner in the third image indicates the spot count using spot (foci) mask. These images are produced using blood cells exposed to 2 Gy γ-irradiation

11. Select wizards from the Guided Analysis menu and choose "Spot Wizard" in the Wizards windows (*see* **Note 14**).

12. Select single cells with the best focus and spot image channel.

13. Select two truth populations – one with high and one with low spot count.

14. Spot wizard generates a mask and features for spot counting based on the truth population images (*see* **Note 15**, Fig. 4).

15. After automatic analysis, a new histogram of the spot count feature for the single cell population is added (*see* **Note 16**, Fig. 5).

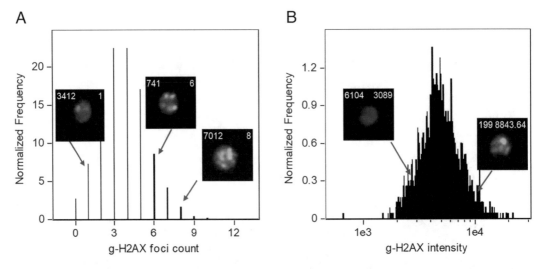

Fig. 5 Histogram of γ-H2AX levels analyzed in the IDEAS software. This graph was generated from blood cells exposed to 2 Gy γ-irradiation. γ-H2AX foci count is calculated based on the spot mask generated through Spot Wizard (**a**). γ-H2AX intensity represents the fluorescence intensity of Ch 2 (γ-H2AX) (**b**). Representative cell images are displayed in the histograms

16. Add statistics of spot count feature (i.e., mean, standard error, or median of spot count) in the Statistics table (*see* **Notes 17–18**).

17. Follow **steps 18–21** to analyze multiple samples at one time (*see* **Note 19**).

18. Select "Batch Data files" in the "Tools" menu.

19. Click "Add Batch" and define a batch (i.e., selecting the files for the batch, compensation matrix, and template file you prepared for γ-H2AX analysis) (.ast or .daf). (*see* **Note 20**).

20. When "Submit Batches" is clicked, the listed files are processed using the template file.

21. After the processing step is complete, the compensation and data files (.cif and .daf) of all the files are generated (*see* **Note 21**). The measurements (γ-H2AX foci number and fluorescence levels) of all the samples are displayed in one Statistics report file.

4   Notes

1. Lithium-heparinized tubes are used here for blood collection because cytogenetic markers as well as γ-H2AX are assessed for radiation biodosimetry. It is possible to use sodium heparin or EDTA vacutainer tubes.

2. Multichannel pipette can be used for high-throughput liquid handling in this method.

3. This method is used to evaluate γ-H2AX levels in human blood samples. The cell culture and sample preparation methods can be different depending on target samples.

4. Blood volume to be collected can be different depending on the study design (i.e., study objectives, no. of replicates, no. of treatment, etc.). In this method, finger-stick-sized blood samples (≤100 μL) are enough to analyze γ-H2AX levels.

5. Blood to medium ratio is critical because cells don't grow well in less medium. We use 1:9 ratio and you need to keep this ratio for culturing higher blood volume.

6. To evaluate the kinetics of γ-H2AX formation for DSB repair capacity, blood samples are needed to be incubated at different times. If assessing time-kinetics of γ-H2AX formation is not needed, skip cell incubation Subheading 3.1, **step 3** and blood aliquots are directly lysed and fixed in Subheading 3.2.

7. The lysis and fixation steps in our method do not require the isolation of peripheral blood mononuclear cells, for example using a Ficoll gradient. Therefore, this could reduce the processing time for performing the γ-H2AX assay using blood samples.

8. If necessary, keep the fixed samples with γ-H2AX (at the same concentration) at 4 °C overnight.

9. We set the 488 nm excitation laser to 200 mW and use it at 40× magnification. The laser power can be adjusted based on Raw Max pixel value of each channel to avoid fluorescence saturation. BF images are captured on channel 1; γ-H2AX immunostaining images, on channel 2; DRAQ5 images, on channel 5; and side scatter, on channel here. The channel designation can be changed by users.

10. We acquire images of 5000–12,000 single cells to obtain around 1000 non-apoptotic lymphocytes, which can be different depending on the users. If you want to adjust the gates for the target cell population in the data analysis step, you can save all the cell images obtained while acquiring enough single cells. In this case, the system should have enough space available to save large files.

11. Samples can be uploaded manually or using the AutoSampler to load samples from multi-well plates. The AutoSampler option could enhance the efficiency and capacity of our high-throughput γ-H2AX assay.

12. To analyze the γ-H2AX levels of all white blood cells or other sub-populations of cells, skip this step or choose other cell populations in a scatter plot based on the area of BF images and SSC.

13. The pan-nuclear γ-H2AX response is known to be a marker for the identification of apoptotic cells among cells with damaged DNA [17, 18]. The measurement of γ-H2AX levels in apoptotic cells may be a confounding factor in the evaluation of radiation-induced DNA damage and individual DNA repair activity [16, 19]. To analyze the radiation-induced changes in γ-H2AX in non-apoptotic cells, we excluded pan-nuclear γ-H2AX-stained cells.

14. You can use the spot wizard tool provided in the IDEAS software. This will create an analysis template including a mask, the set of pixels containing the region of interest, and features for counting γ-H2AX foci. You can also prepare the analysis template manually using the feature finder and mask manager.

15. You can find newly generated mask definitions in a mask manager. Based on our truth populations, we obtained mask "Range" (Peak(Spot(M02, Channel 2, Bright, 4.5, 1), Channel 2, Bright,0), 0–200,0–1) for γ-H2AX foci count. Researchers should prepare their best mask to reflect the different protocols and environment of each lab.

16. You need to check the performance of the mask for other cell images. If necessary, adjust the gates for the target cell population and two truth populations and repeat the wizard after viewing the images and validating the spot counts. You can also adjust the analysis template manually.

17. If you are interested in determining the fluorescence intensity and count of foci of γ-H2AX, you can create features on spot fluorescence and add related values in the Statistics Table. Select the "Features" option in the "Analysis" menu. You can create a new feature for the γ-H2AX fluorescence by selecting the appropriate feature type and mask (i.e., "Intensity" as feature type, γ-H2AX spot mask).

18. You can export the results of statistical analysis as the Statistics Report (.txt). However, you need to define the Statistics Report file first. Select the "Define Statistics Report" in the "Reports" menu. You can input the statistical data of target populations (i.e., count, %Gated, mean of γ-H2AX measurements) and generate the report.

19. Batch processing allows the automatic analysis of a group of files with one template. This process can reduce human error in the counting of γ-H2AX foci, providing accurate and reliable results.

20. For batch processing, you need to prepare a template file first. The data file used to analyze γ-H2AX levels in Step 3.4.16 can be used as the template file. If you define the Statistics Report in the template file, you will receive a Statistics Report file for all your samples.

21. You can check the characteristics (cell population, γ-H2AX levels, single cell images, etc.) of a sample in each data file. If the current γ-H2AX analysis procedure seems to be inappropriate for application in other samples, the template file can be modified and all samples should be re-run.

Acknowledgements

This study was supported by a grant of the Korea Institute of Radiological and Medical Sciences, funded by Nuclear Safety and Security Commission of the Republic of Korea (50091-2023). This work was also supported by the Center for High-Throughput Minimally-Invasive Radiation Biodosimetry, National Institute of Allergy and Infectious Diseases (grant number U19AI067773).

References

1. Rothkamm K, Horn S (2009) Gamma-H2AX as protein biomarker for radiation exposure. Annali dell'Istituto superiore di sanita 45:265–271

2. Rogakou EP, Pilch DR, Orr AH, Ivanova VS, Bonner WM (1998) DNA double-stranded breaks induce histone H2AX phosphorylation on serine 139. J Biol Chem 273:5858–5868

3. Mah LJ, El-Osta A, Karagiannis TC (2010) gammaH2AX: a sensitive molecular marker of DNA damage and repair. Leukemia 24:679–686

4. Dickey JS, Redon CE, Nakamura AJ, Baird BJ, Sedelnikova OA, Bonner WM (2009) H2AX: functional roles and potential applications. Chromosoma 118:683–692

5. Moquet J, Barnard S, Staynova A, Lindholm C, Monteiro Gil O, Martins V, Rossler U, Vral A, Vandevoorde C, Wojewodzka M, Rothkamm K (2017) The second gamma-H2AX assay intercomparison exercise carried out in the framework of the European biodosimetry network (RENEB). Int J Radiat Biol 93:58–64

6. Roch-Lefevre S, Mandina T, Voisin P, Gaetan G, Mesa JE, Valente M, Bonnesoeur P, Garcia O, Voisin P, Roy L (2010) Quantification of gamma-H2AX foci in human lymphocytes: a method for biological dosimetry after ionizing radiation exposure. Radiation Res 174:185–194

7. Horn S, Barnard S, Rothkamm K (2011) Gamma-H2AX-based dose estimation for whole and partial body radiation exposure. PLoS One 6:e25113

8. Moquet J, Barnard S, Rothkamm K (2014) Gamma-H2AX biodosimetry for use in large scale radiation incidents: comparison of a rapid '96 well lyse/fix' protocol with a routine method. Peer J 2:e282

9. Rothkamm K, Lobrich M (2003) Evidence for a lack of DNA double-strand break repair in human cells exposed to very low x-ray doses. Proc Natl Acad Sci U S A 100:5057–5062

10. Muslimovic A, Ismail IH, Gao Y, Hammarsten O (2008) An optimized method for measurement of gamma-H2AX in blood mononuclear and cultured cells. Nat Protoc 3:1187–1193

11. Johansson P, Fasth A, Ek T, Hammarsten O (2017) Validation of a flow cytometry-based detection of gamma-H2AX, to measure DNA damage for clinical applications. Cytometry B Clin Cytom 92:534–540

12. Hamasaki K, Imai K, Nakachi K, Takahashi N, Kodama Y, Kusunoki Y (2007) Short-term culture and gammaH2AX flow cytometry determine differences in individual radiosensitivity in human peripheral T lymphocytes. Environ Mol Mutagen 48:38–47

13. Basiji DA, Ortyn WE, Liang L, Venkatachalam V, Morrissey P (2007) Cellular image analysis and imaging by flow cytometry. Clin Lab Med 27:653–670, viii

14. Basiji D, O'Gorman MR (2015) Imaging flow cytometry. J Immunol Methods 423:1–2

15. Basiji DA (2016) Principles of amnis imaging flow cytometry. Methods Mol Biol 1389:13–21

16. Lee Y, Wang Q, Shuryak I, Brenner DJ, Turner HC (2019) Development of a high-throughput gamma-H2AX assay based on imaging flow cytometry. Radiat Oncol 14:150

17. Ding D, Zhang Y, Wang J, Zhang X, Gao Y, Yin L, Li Q, Li J, Chen H (2016) Induction and inhibition of the pan-nuclear gamma-H2AX response in resting human peripheral blood lymphocytes after X-ray irradiation. Cell Death Discov 2:16011

18. Solier S, Pommier Y (2009) The apoptotic ring: a novel entity with phosphorylated histones H2AX and H2B and activated DNA damage response kinases. Cell Cycle 8:1853–1859

19. Turner HC, Shuryak I, Taveras M, Bertucci A, Perrier JR, Chen C, Elliston CD, Johnson GW, Smilenov LB, Amundson SA, Brenner DJ (2015) Effect of dose rate on residual gamma-H2AX levels and frequency of micronuclei in X-irradiated mouse lymphocytes. Radiat Res 183:315–324

Chapter 8

Label-Free Identification of Persistent Particles in Association with Primary Immune Cells by Imaging Flow Cytometry

Bradley Vis, Jonathan J. Powell, and Rachel E. Hewitt

Abstract

The frequency of human exposure to persistent particles via consumer products, air pollution, and work environments is a modern-day hazard and an active area of research. Particle density and crystallinity, which often dictate their persistence in biological systems, are associated with strong light absorption and reflectance. These attributes allow several persistent particle types to be identified without the use of additional labels using laser light-based techniques such as microscopy, flow cytometry, and imaging flow cytometry. This form of identification allows the direct analysis of environmental persistent particles in association with biological samples after in vivo studies and real-life exposures. Microscopy and imaging flow cytometry have progressed with computing capabilities and fully quantitative imaging techniques can now plausibly detail the interactions and effects of micron and nano-sized particles with primary cells and tissues. This chapter summarises studies which have utilized the strong light absorption and reflectance characteristics of particles for their detection in biological specimens. This is followed by the description of methods for the analysis of whole blood samples and the use of imaging flow cytometry to identify particles in association with primary peripheral blood phagocytic cells, using brightfield and darkfield parameters.

Key words Nanoparticle, Microparticle, Whole blood, Neutrophils, Imaging flow cytometry, Label-free imaging, Persistent particles

1 Introduction

Persistent particles can be defined as particulate matter that is not easily digested, dissolved, or otherwise removed within a biological system. In the environment of today, most of us are continuously exposed to significantly higher amounts of non-biological, persistent particulates compared to our ancestors. Increases in exposure occur through the addition of particles to consumer products [1], environmentally, through air pollution [2], and through occupational exposures [3]. Though much has been uncovered on particle interactions within biological systems, the full extent of particle

Natasha S. Barteneva and Ivan A. Vorobjev (eds.), *Spectral and Imaging Cytometry: Methods and Protocols*, Methods in Molecular Biology, vol. 2635, https://doi.org/10.1007/978-1-0716-3020-4_8, © Springer Science+Business Media, LLC, part of Springer Nature 2023

impact remains unclear, especially at the cellular level [4]. The same dense and crystalline characteristics that tend to facilitate particle persistence in biological environments are often associated with the characteristics of strong light absorption and reflectance, allowing for their identification in association with cells without the use of additional labels using laser light-based techniques (Fig. 1). Increases in reflectance caused by persistent particles appear distinct from the small increases associated with dead or dying cells which have lost membrane integrity (Fig. 2).

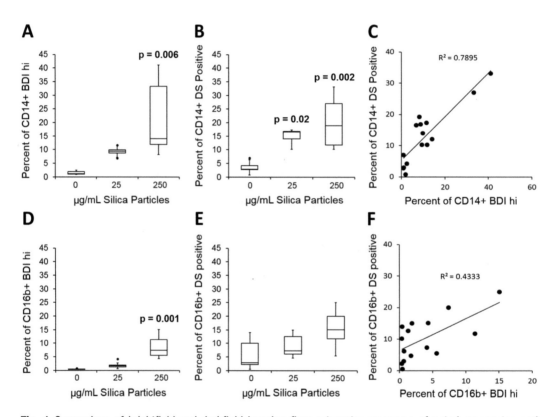

Fig. 1 Comparison of brightfield and darkfield imaging flow cytometry measures of gated monocytes and neutrophils after crystalline silica exposure. Boxplots show the percentage of CD14$^+$ monocytic cells gated as (**a**). Bright detail intensity (BDI) high in darkfield or (**b**) dark spot (DS) positive in brightfield in response to increasing dose of crystalline silica in the whole blood of $n = 5$ individual subjects. (**c**) Correlation between CD14$^+$ cells gated as darkfield BDI high and darkspot (DS) in brightfield positive, Pearson correlation $r = 0.888$, $p = <0.01$. Percentage of CD16b$^+$ neutrophils gated as (**d**). BDI high in darkfield or (**e**). DS positive in brightfield in response to increasing dose of crystalline silica in whole blood of $n = 5$ subjects. (**f**) Correlation between CD16b$^+$ cells gated as BDI high and dark spot (DS) positive, Pearson correlation $r = 0.658$, $p \leq 0.05$ Boxplots display Q1-Q3 with whiskers set at 1.5 × IQR (interquartile range) above the third quartile and 1.5 × IQR below the first quartile, minimum or maximum values that have fallen outside this range are shown as outliers (small black dots). Significant difference from the control was assessed using Tukey's honest significance test for multiple comparisons following a one-way ANOVA, Tukey's honestly significant difference (HSD) p-values where significant are shown

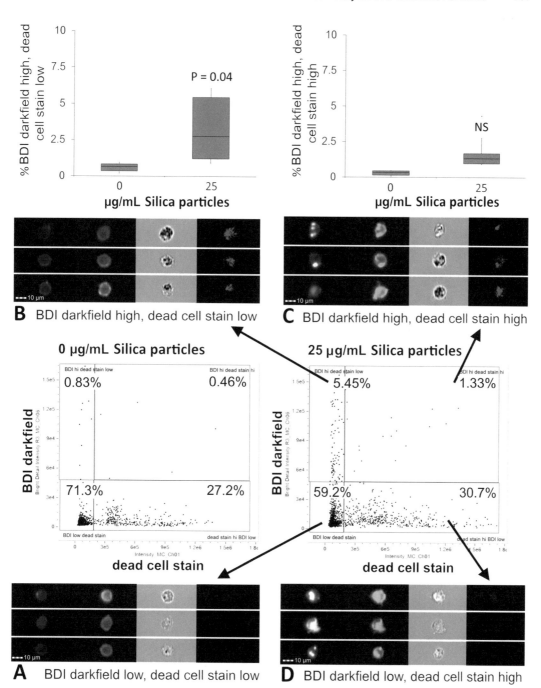

Fig. 2 Darkfield bright detail intensity fluorescent versus dead cell staining in cells after incubation with particulate crystalline silica. Example imaging flow cytometry dot plots with bright detail intensity (BDI) of darkfield spots on the vertical axis and dead cell stain intensity on the horizontal axis. Quadrants applied to the dot plot reveals percentages of cells falling within (**a**) the double negative section of the quadrant, (**b**) cells single positive for increased darkfield BDI, (**c**) double-positive cells and (**d**) cells single positive for the dead cell stain. Representative cell images for each quadrant section are shown. Boxplots directly above the associated example cell images show the percentage of CD14$^+$ cells residing in the BDI darkfield high and dead stain low section of the quadrant (**b**), or BDI darkfield high and dead stain high quadrant (**c**) in response to increasing dose of crystalline silica in the whole blood of $n = 5$ individual subjects. The difference from the 0 μg/mL particle control incubation was measured using a paired, two-tailed Student's T-test

Table 1
Particulates identified by strong light absorption and reflectance characteristics

Particle	Technique used	Reference
Gold (Au)	Flow cytometry, microscopy	[13, 14] [21]
Silver (Ag)	Flow cytometry, microscopy	[5, 8, 15] [15]
Titanium dioxide (TiO$_2$)	Flow cytometry, microscopy, imaging flow cytometry	[5–9] [10, 11] [9, 12]
Iron oxide (Fe$_3$O$_4$)	Flow cytometry	[5, 8]
Copper oxide (CuO)	Flow cytometry	[8]
Zinc oxide (ZnO)	Flow cytometry	[8]
Monosodium urate (NaC$_5$H$_3$N$_4$O$_3$·H$_2$O)	Flow cytometry, microscopy	[22]
Crystalline silica (SiO$_2$)	Flow cytometry, microscopy, imaging flow cytometry	[18–20] [16, 17] [20]
Carbon nanotubes (C)	Imaging flow cytometry	[23]

A summary of particles successfully identified by these characteristics in association with cells is shown in Table 1. For example, pigment-grade titanium dioxide (TiO$_2$) particles have been distinguished by their light scatter features by conventional flow cytometry [5–9], microscopy [10, 11], and more recently, imaging flow cytometry [9, 12]. Particles of gold (Au), silver (Ag), iron oxide (Fe$_3$O$_4$), copper oxide (CuO), and zinc oxide (ZnO) have all been measured in association with cells by flow cytometry [5, 8, 13–15]. Crystalline silica has been detected by microscopy [16, 17], conventional flow cytometry [18, 19], and imaging flow cytometry [20]. It is important to note that these methods do not readily distinguish between the materials which have a high capacity to absorb or scatter light but rather allow the particles to be distinguished from the cell structures themselves. Despite this limitation, label-free identification allows for the direct ex-vivo analysis of persistent particles after in vivo studies and real-life exposures. Additionally, label-free identification circumvents a plethora of problems associated with significantly altered protein corona formation on the surface of particles as a result of particle processing and fluorescent labeling, compared to protein coronas formed in the absence of these labels [24, 25]. This is particularly pertinent to studying the effects of persistent particle exposures.

The fields of microscopy and imaging flow cytometry have progressed in line with advances in computing power, and fully quantitative imaging techniques for detailing the interactions and effects of micron and nano-sized particles with primary cells and tissues are feasible. Imaging flow cytometry merges conventional flow cytometry throughputs with microscopic imaging making it a helpful technique for examining particle-cell events in detail [9, 12, 23, 26, 27]. Advances integrating machine learning with image analysis interfaces have created new and powerful ways of analyzing multiple images in a fully quantitative fashion, transforming the information gained from image data. Nano and sub-micron-sized persistent particulate materials can interact with cells in many ways, with a diversity of outcomes [28]. As such, quantitate image-based information has the potential to assist in developing a complete understanding of particulate interactions at the cellular level, especially relating to the fields of immunology and toxicology. Advances in imaging analysis allow small details within the images, often missed by the naked eye, however, to be measured and quantified in an autonomous fashion.

In this chapter, we describe the use of an Imagestream multispectral flow cytometer (Amnis- Luminex) to focus on the technique of using brightfield (where light-absorbing materials appear as dark regions) and darkfield parameters (where reflectance or high light-scattering properties are identified), thereby avoiding the use of fluorescent labeling, to identify persistent particles (Fig. 1). It should be noted that the principles described could be applied to a variety of other cell imaging platforms (Table 1). Freshly drawn peripheral blood was used as our cell-rich biological matrix. CD14$^+$ monocytes and CD16b$^+$ neutrophils were our target cells as the predominant phagocytes within the fresh whole blood, and crystalline silica particles were utilized as an exemplar persistent particle. Detailed analysis of particle-associated darkfield fluorescence, combined with dead cell staining (Fig. 2), also demonstrates that increases in darkfield fluorescence associated with reflective particles appear distinct from small increases associated with dead or dying cells which have lost membrane integrity [29].

2 Materials

1. Vacutainer blood collection tubes containing heparin (*see* **Note 1**).

2. A small shaker, rocker or rotator (*see* **Note 2**).

3. Sodium heparin stock solution was filtered using a 0.2-micron membrane filter unit and stored at 2-8 °C.

4. Nano or sub-micron-sized particles of interest. (*see* **Note 3**).

5. Red Blood cell lysis buffer. A commercial 10× ammonium chloride-based lysing reagent BD Pharm Lyse (BD Biosciences).

6. Wash buffer. Sterile 2% bovine serum albumin (BSA) in phosphate-buffered saline (PBS). Stored at 4 °C and kept on ice whilst in use.

7. PBS 1×.

8. 2% Formaldehyde solution in sterile PBS.

9. Fluorescently conjugated antibody staining cocktail. Titrated Alexa 488 conjugated anti-human CD14 for the identification of monocytes and PE-conjugated anti-human CD16b to identify neutrophils.

10. Live/dead violet fluorescent reactive fixable dye (Invitrogen #L34955) (*see* **Note 4**).

11. Sterile Tubes. 15 and 50 mL conical polypropylene tubes; 5 mL round bottom polystyrene test tubes with snap cap, and 5 mL round bottom polystyrene test tubes with 35 μm nylon mesh cell strainer caps.

12. Imaging flow cytometer Imagestream X (Amnis-Luminex) equipped with 405, 488 nm excitation lasers and 785 nm laser for scatter signal, INSPIRE acquisition software, and IDEAS analysis software (Amnis-Luminex).

3 Methods

3.1 Blood Collection and Particle Incubation

1. Collect blood by venepuncture from healthy donors following informed consent into vacutainer tubes containing lithium-heparin. Working within a sterile Class II biological safety cabinet, promptly transfer whole blood (WB) to a sterile conical polypropylene tube and add heparin solution for a final concentration of 0.5 mg heparin per mL of WB.

2. Aliquot WB (up to 200 μL) into sterile 5 mL round bottom polystyrene test tubes, treat with particles (in the example shown, 0–250 μg/mL crystalline silica particles were used), and incubate at 37 °C, with 5% CO_2, for the desired duration (anywhere between 30 min and 24 h) on a gentle shaker to prevent settling of cells and particles and to prevent blood separation. At this time it is important to set up unchallenged cells as a negative control for each subject, as well as compensation tubes for each stain and cells plus particle (at the highest concentration) only sample for compensation and analysis alongside the main experiment.

3.2 Red Blood Cell Lysis

1. Working within a sterile Class II biological safety cabinet, prepare the 1× red blood cell (RBC) lysis solution by diluting BD Pharm Lyse with sterile tissue culture-grade water. Ensure solution is at room temperature, according to manufacturers' instructions, prior to use.

2. To lyse RBC, add 2 mL 1× Pharm Lyse solution to each sample tube containing up to 200 μL WB and immediately vortex gently.

3. Incubate for 15 min at room temperature in the dark.

4. Transfer tubes to a centrifuge and spin at 200g for 5 min.

5. Gently aspirate the supernatants taking care not to disrupt pelleted cells.

6. Top up the tubes with PBS to wash cells and centrifuge to pellet cells at 200g for 5 min, and gently aspirate the supernatants. Repeat twice.

7. Top up the tubes with PBS and centrifuge at 1500 rpm for 5 min.

3.3 Dead Cell Staining

1. Carefully decant and blot away the PBS wash supernatant leaving only the pelleted cells within the tubes.

2. Prepare the live live-dead stain according to the manufacturer's specifications.

3. Resuspend the cells in the stain and incubate according to the manufacturer's specifications.

4. Stain the experimental samples/controls and also the single stain live/dead compensation tube at this time.

5. Wash with cold wash buffer, centrifuging at 1500 rpm for 5 min.

3.4 Phenotypic Marker Surface Staining

1. Decant wash buffer supernatant and resuspend cells in the residual buffer (around 200 μL).

2. Add the cell surface antibody cocktail to the experimental samples/controls and incubate according to the manufacturer's specifications, incubate on ice in the dark for at least 20 min.

3. Add single stains using the same staining volumes to the single stain compensation tubes for the cell surface markers used (image analysis controls) at this time and treat these tubes in the same way as the experimental sample/control tubes.

4. After staining, wash cells again with cold wash buffer re-suspend in a small volume of PBS containing 2% formaldehyde.

5. Store cell samples on ice and in the dark and acquire within hours.

3.5 Acquisition Using an ImageStreamX Platform. Instrument Set up

For the experimental methods described, imaging flow cytometry acquisition and analysis were carried out using an ImagesStream X platform (Amnis-Luminex). It should be noted that the principles of these methods could be applied to other imaging and analysis platforms. Alternative fluorophores may be used for different configurations of instruments. During the sample staining steps, the Imagestream fluidics should be initialized, followed by the running of all calibration and test scripts using INSPIRE software (Amnis-Luminex). All cell samples were filtered through 35 µm nylon cell strainer mesh tubes (BD Biosciences, UK) immediately before acquisition.

1. Begin by running the cells plus sample containing only particles to define cell events and exclude cell clumps and debris (*see* **Note 5**). On an Imagestream X, this is achieved by setting the area upper and lower limits of the channel collecting brightfield images within the cell classifier.

2. Use the cells plus particle sample to check the scatter (785 nm) laser power, ensuring that events are not saturating the camera for the darkfield (scatter) channel. Saturated events appear on the axis of the Raw Max Pixel plots at a value of 4095. Set the laser power to remove any saturated events.

3. Repeat the process using the single stained compensation tubes to set the remaining laser power. Attenuate the laser power so that each sample's positive signal can be easily distinguished from negatively staining cells but to be not so high as to result in saturated events in the appropriate channels.

4. Run the fully stained samples, collecting sufficient events to allow for any target populations which appear at lower frequencies within the mixed cell population of whole blood (i.e. monocytes, compared to neutrophils). For the data shown, 30,000 events were collected per sample.

5. Turn off the brightfield, 785 nm laser, and the cell classifier to run the single stained compensation tubes, collecting 500–1000 bright events for each stain used for the generation of compensation matrices in the analysis step.

3.6 Acquisition Using an ImageStreamX Platform. Spectral Overlap Compensation

1. The start of analyzing a new data file within the IDEAS software prompts the generation of the compensation matrix, consisting of the single stained compensation files acquired simultaneously as the experimental samples.

2. Once complete, it is important to check for indicators of over- (black holes in images) and under-compensation (fluorescent leakage into adjacent channels) using the image gallery and editing the compensation matrix as necessary.

3. Once generated, the same compensation matrix should be applied to all the experimental samples acquired.

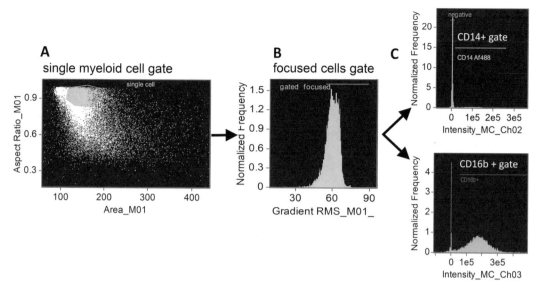

Fig. 3 Initial gating analysis strategy for the identification of crystalline silica cellular association using brightfield and darkfield parameters in imaging cytometry. (**a**) Area (size of the masked cells in μm^2) versus aspect ratio (of the minor axis divided by the major axis) of the brightfield cell images were used to draw an initial dot plot to identify cells of interest and exclude doublets and debris. From the single cell (phagocyte) gate cells in best focus were gated (using gradient RMS, shown in (**b**)), followed by gating on CD16b$^+$ or CD14$^+$ positives using fluorescence intensity (histograms for CD16b$^+$ in Channel 3, CD14$^+$ in Channel 2) shown in (**c**)

3.6.1 Data Analysis. Initial Gating for Single, Focused, Cells Positive for Phenotypic Markers

1. Use brightfield area versus brightfield aspect ratio (named the 'single-cell default' in IDEAS) to create a dot plot to identify and gate the single myeloid cell population by clicking on dots within the plot and examining the corresponding cell images (Fig. 3a).

2. Next, create a gradient RMS histogram plot of the cells within the single-cell gate using the building blocks and gate on cells in the best focus (Fig. 3b).

3. Use the gated, single, best-focused cells to create further fluorescence intensity histograms of phenotypic markers using the focused, single gated cells plots (Alexa-488-CD14$^+$ monocytes in channel 2PE-CD16b$^+$ neutrophils in channel 3 in this example), and gate on the positive populations (Fig. 3c).

3.6.2 Data Analysis. Measuring Particle Association with Cells with Darkfield Image Analysis

1. Using the single, focused, fluorescent positive gated cells (e.g., single, focused, Alexa-488-CD14$^+$), create a further dot plot of the bright detail intensity R3 (BDI R3) in darkfield on the horizontal axis and brightfield bright detail intensity on the vertical axis (*see* **Note 6**).

2. Establish where the natural amount of darkfield signal for the cells resides by analyzing the untreated control first. Place the

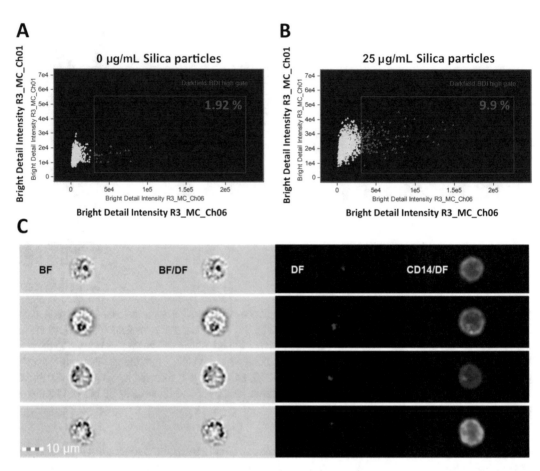

Fig. 4 Bright detail intensity in darkfield measurements. Single, focused, gated CD14⁺ population of cells obtained from the initial gating (shown in Fig. 3c) plotted as bivariate plots using bright detail intensity (BDI) measurements for both brightfield (vertical axis) and darkfield (horizontal axis). (**a**) A region selecting cells displaying increases in darkfield (Channel 6) BDI was then drawn based on the 0 μg/mL particle control for each subject, and applied to particle exposed cells. (**b**) Example analysis images of CD14⁺ (green) cells residing within the DF-BDI high gate (darkfield BDI displayed in pink) are shown in (**c**), examples were selected that additionally display good localisation of the crystalline silica particles in brightfield (BF) as dark spots. Displayed from left to right; Brightfield (BF); brighfield/darkfield merged image (BF/DF); darkfield (DF) and CD14/DF merged image

darkfield BDI positive gate beyond the natural population cluster for BDI-R3 (shown in Fig. 4). Once optimized, the same gate should be applied to all samples from the same donor. The process should be repeated for each donor as significant variations occur between donors in terms of cell populations, receptor expression fluorescence intensities, and even darkfield fluorescence of cells from different donors (see **Note 7**).

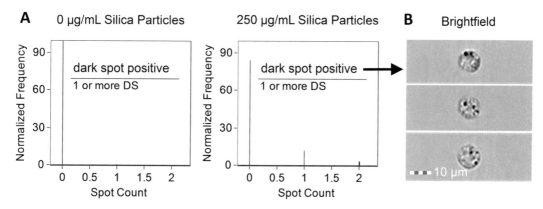

Fig. 5 Spot count measurement of dark spots in brightfield images Single, focused, gated CD16b⁺ population of cells obtained from the initial gating (shown in Fig. 3c) were utilised for spot counts, measuring dark spots appearing within the brightfield images, identified and quantified using the spot count feature. Dark spot counts of the 0 µg/mL and 250 µg/mL crystalline silica incubations are shown in (**a**), alongside example brightfield images positive for dark spots in (**b**)

3.7 Measurement of Particle Cell Association Using Brightfield Images

1. Use the mask manager to create a cell mask in the brightfield channel, applying the peak, dark, mask feature and with care adjust the spot-to-cell ratio until the masked area is located over the darkest spots. Test the ratio by selecting positive images from the object gallery within the mask manager until the mask is optimized.

2. Add a spot mask feature, adjust the spot-to-cell background ratio, and radius values, until it convincingly covers only punctate dark spot areas.

3. To use the newly made mask, create a new feature in the feature manager. Select the spot count feature and the optimized spot count mask to create the feature. Once the new spot count feature is created, use the single, focused, fluorescent positive gated cells (Fig. 3c) to create a spot count histogram, selecting your newly made spot count feature on the vertical axis. Examples of spot counts performed using the single, focused, CD16b⁺ positive neutrophils are shown in Fig. 5.

4 Notes

1. A study by Baumann et al. [30] compared the use of anticoagulants commonly used in blood collection tubes and found coagulants other than heparin (sodium citrate or EDTA) hindering particle cell interactions in whole blood incubations. For this reason, heparin was used as an anticoagulant throughout the whole blood incubation steps.

2. Required for the gentle agitation of the blood samples to prevent blood separation within the incubator during the particle incubation step.

3. Preferably either tissue culture grade or heat-treated to break down any immunogenic contaminating components, suspended in sterile water and sized prior to adding to whole blood or cells. Preparation and size characterization of particles: For particles that cannot be sourced as sterile tissue culture-grade powders, heat treatment to break down any contaminating bacterial components is required so as not to alter the cellular response via bacterial contamination and immune activation. As described by others [31], this was achieved by baking the particles in powder form at 200 °C prior to suspending them in sterile H_2O. The resulting particle suspensions should be appropriately characterized prior to their use. For the data described, we used Min-u-sil 5 quartz silica particles (US Silica Company); a Malvern Mastersizer 2000 (Malvern Instruments), using parameters specific for silicon oxide (refractive index = 1.45, absorption = 0.01). Replicate measurements were recorded, and the volume distribution was assessed.

4. Violet fluorescent reactive fixable dye is a commercial fluorescent reactive dye which reacts with cellular proteins (amines). These dyes cannot penetrate live cell membranes, so only dead or dying cells with damaged membranes fluoresce brightly upon staining.

5. Although not a critical step, taking time to optimize the area upper and lower limits for your cell populations of interest significantly refines the number of cells of interest in saved data files. This experiment targeted phagocytic cells with the area upper limit (AUL) set to 600 and the area lower limit (ALL) set to 50 units in the brightfield channel.

6. Measures other than brightfield and bright detail intensity could be used here on the vertical axis.

7. It is essential to check and adjust gates by examining the corresponding images in each set of the analysis. Inspection of the analysis masks and adjusting as needed also greatly assists in the generation of accurate analyses. For more detail on the successful generation and use of masks in imaging flow cytometry see Dominical et al. [32].

Acknowledgements

The authors wish to thank the Volunteer Studies and Clinical Services Team at MRC EWL for recruitment and consent of volunteers and the volunteers' blood collection and the UK Medical Research Council for their support (Grant number MR/R005699/1).

References

1. Larsen PB, Christensen F, Jensen KA, Brinch A, Mikkelsen SH (2015) Exposure assessment of nanomaterials in consumer products. Danish Environmental Protection Agency Environmental Project 1636. https://www2.mst.dk/Udgiv/publications/2015/01/978-87-93283-57-2.pdf. Accessed 05/11/2019

2. Künzli N, Kaiser R, Medina S, Studnicka M, Chanel O, Filliger P, Herry M, Horak F Jr, Puybonnieux-Texier V, Quénel P, Schneider J, Seethaler R, Vergnaud JC, Sommer H (2000) Public-health impact of outdoor and traffic-related air pollution: a European assessment. Lancet 356:795–801. https://doi.org/10.1016/S0140-6736(00)02653-2

3. Baldwin PEJ, Yates T, Beattie H, Keen C, Warren N (2019) Exposure to respirable crystalline silica in the GB brick manufacturing and stone working industries. Ann Work Expo Health 63:184–196. https://doi.org/10.1093/annweh/wxy103

4. Riediker M, Zink D, Kreyling W, Oberdörster G, Elder A, Graham U, Lynch I, Duschl A, Ichihara G, Ichihara S, Kobayashi T, Hisanaga N, Umezawa M, Cheng TJ, Handy R, Gulumian M, Tinkle S, Cassee F (2019) Particle toxicology and health – where are we? Part Fibre Toxicol 16:–19. https://doi.org/10.1186/s12989-019-0302-8. Erratum in: Part Fibre Toxicol. 16:26

5. Suzuki H, Toyooka T, Ibuki Y (2007) Simple and easy method to evaluate uptake potential of nanoparticles in mammalian cells using a flow cytometric light scatter analysis. Environ Sci Technol 48:3018–3024. https://doi.org/10.1021/es0625632

6. Ashwood P, Thompson RP, Powell JJ (2007) Fine particles that adsorb lipopolysaccharide via bridging calcium cations may mimic bacterial pathogenicity towards cells. Exp Biol Med 232:107–117

7. Zucker RM, Massaro EJ, Sanders KM, Degn LL, Boyes WK (2010) Detection of TiO$_2$ nanoparticles in cells by flow cytometry. Cytometry A 77A:677–685

8. Toduka Y, Toyooka T, Ibuki Y (2012) Flow cytometric evaluation of nanoparticles using side-scattered light and reactive oxygen species-mediated fluorescence–correlation with genotoxicity. Environ Sci Technol 46:7629–7636

9. Hewitt RE, Vis B, Pele LC, Faria N, Powell JJ (2017) Imaging flow cytometry assays for quantifying pigment grade titanium dioxide particle internalization and interactions with immune cells in whole blood. Cytometry A 91:1009–1020

10. Thoree V, Skepper J, Deere H, Pele LC, Thompson RP, Powell JJ (2008) Phenotype of exogenous microparticle-containing pigment cells of the human Peyer's patch in inflamed and normal ileum. Inflamm Res 57:374–378. https://doi.org/10.1007/s00011-007-7216-x

11. Gibbs-Flournoy EA, Bromberg PA, Hofer TP, Samet JM, Zucker RM (2011) Darkfield-confocal microscopy detection of nanoscale particle internalization by human lung cells. Part Fibre Toxicol 8:2

12. Vranic S, Boggetto N, Contremoulins V, Mornet S, Reinhardt N, Marano F, Baeza-Squiban A, Boland S (2013) Deciphering the mechanisms of cellular uptake of engineered nanoparticles by accurate evaluation of internalization using imaging flow cytometry. Part Fibre Toxicol 10:2

13. Park J, Ha MK, Yang N, Yoon TH (2017) Flow cytometry-based quantification of cellular Au nanoparticles. Analytical Chem 84:2449–2456

14. Wu Y, Ali MR, Dansby K, El-Sayed MA (2019) Improving the flow cytometry-based detection of the cellular uptake of gold nanoparticles. Analytical Chem 91:14261–14267

15. Zucker RM, Ortenzio J, Degn LL, Lerner JM, Boyes WK (2019) Biophysical comparison of four silver nanoparticles coatings using microscopy, hyperspectral imaging and flow cytometry. PLoS One 14:e0219078

16. Tian L, Dai S, Wang J, Huang Y, Ho SC, Zhou Y, Lucas D, Koshland CP (2008) Nano-quartz in Late Permian C1 coal and the high incidence of female lung cancer in the Pearl River Origin area: a retrospective cohort study. BMC Public Health 8:398

17. Hornung V, Bauernfeind F, Halle A, Samstad EO, Kono H, Rock KL, Fitzgerald KA, Latz E (2008) Silica crystals and aluminum salts activate the NALP3 inflammasome through

phagosomal destabilization. Nat Immunol 9:847–856

18. Beamer CA, Holian A (2005) Scavenger receptor class A type I/II (CD204) null mice fail to develop fibrosis following silica exposure. Am J Physiol Lung Cell Mol Physiol 289:L186–L195

19. Beamer GL, Seaver BP, Jessop F, Shepherd DM, Beamer CA (2016) Acute exposure to crystalline silica reduces macrophage activation in response to bacterial lipoproteins. Front Immunol 7:49

20. Vis B, Powell JJ, Hewitt RE (2020) Imaging flow cytometry methods for quantitative analysis of label-free crystalline silica particle interactions with immune cells. AIMS Biophys 7:144

21. Chithrani BD, Ghazani AA, Chan WC (2006) Determining the size and shape dependence of gold nanoparticle uptake into mammalian cells. Nano Lett 6:662–668

22. Schorn C, Janko C, Latzko M, Chaurio R, Schett G, Herrmann M (2012) Monosodium urate crystals induce extracellular DNA traps in neutrophils, eosinophils, and basophils but not in mononuclear cells. Front Immunol 3:277

23. Marangon I, Boggetto N, Ménard-Moyon C, Luciani N, Wilhelm C, Bianco A, Gazeau F (2013) Localization and relative quantification of carbon nanotubes in cells with multispectral imaging flow cytometry. J Vis Exp 82:e50566

24. Neagu M, Piperigkou Z, Karamanou K, Engin AB, Docea AO, Constantin C, Negrei C, Nikitovic D, Tsatsakis A (2017) Protein bio-corona: critical issue in immune nanotoxicology. Arch Toxicol 91:1031–1048

25. Zanganeh S, Spitler R, Erfanzadeh M, Alkilany AM, Mahmoudi M (2016) Protein corona: opportunities and challenges. Int J Biochem Cell Biol 75:143–147

26. Phanse Y, Ramer-Tait AE, Friend SL, Carrillo-Conde B, Lueth P, Oster CJ, Phillips GJ (2012) Analyzing cellular internalization of nanoparticles and bacteria by multi-spectral imaging flow cytometry. J Vis Exp 64:e3884

27. Smirnov A, Solga MD, Lannigan J, Criss AK (2015) An improved method for differentiating cell-bound from internalized particles by imaging flow cytometry. J Immunol Methods 423:60–69

28. Hewitt RE, Chappell HF, Powell JJ (2020) Small and dangerous? Potential toxicity mechanisms of common exposure particles and nanoparticles. Current Opinion in Toxicol 19:93–98

29. George TC, Basiji DA, Hall BE, Lynch DH, Ortyn WE, Perry DJ, Seo MJ, Zimmerman CA, Morrissey PJ (2004) Distinguishing modes of cell death using the ImageStream multispectral imaging flow cytometer. Cytometry A 59:237–245

30. Baumann D, Hofmann D, Nullmeier S, Panther P, Dietze C, Musyanovych A, Ritz S, Landfester K, Mailänder V (2013) Complex encounters: nanoparticles in whole blood and their uptake into different types of white blood cells. Nanomedicine 8:699–713

31. Satpathy SR, Jala VR, Bodduluri SR, Krishnan E, Hegde B, Hoyle GW, Fraig M, Luster AD, Haribabu B (2015) Crystalline silica-induced leukotriene B4-dependent inflammation promotes lung tumour growth. Nat Commun 6:7064

32. Dominical V, Samsel L, McCoy JP Jr (2017) Masks in imaging flow cytometry. Methods 112:9–17

Chapter 9

"Immuno-FlowFISH": Applications for Chronic Lymphocytic Leukemia

Henry Y. L. Hui, Wendy N. Erber, and Kathy A. Fuller

Abstract

Imaging flow cytometry has the capacity to bridge the gap that currently exists between the diagnostic tests that detect important phenotypic and genetic changes in the clinical assessment of leukemia and other hematological malignancies or blood-based disorders. We have developed an "Immuno-flowFISH" method that leverages the quantitative and multi-parametric power of imaging flow cytometry to push the limits of single-cell analysis. Immuno-flowFISH has been fully optimized to detect clinically significant numerical and structural chromosomal abnormalities (i.e., trisomy 12 and del(17p)) within clonal CD19/ CD5+ CD3− Chronic Lymphocytic Leukemia (CLL) cells in a single test. This integrated methodology has greater accuracy and precision than standard fluorescence in situ hybridization (FISH). We have detailed this immuno-flowFISH application with a carefully catalogued workflow, technical instructions, and a repertoire of quality control considerations to supplement the analysis of CLL. This next-generation imaging flow cytometry protocol may provide unique advancements and opportunities in the holistic cellular assessment of disease for both research and clinical laboratory settings.

Key words Imaging flow cytometry, Fluorescence in situ hybridization, Chronic lymphocytic leukemia

1 Introduction

In this chapter, we describe a new methodology for the analysis of chromosomes in cells identified by their phenotype. This new cytogenomic method has been called "Immuno-flowFISH", because of the combination of "immunophenotype," fluorescence in situ hybridization (*FISH*), and the use of flow cytometry. The incorporation of cell phenotype ensues and only the specific cells of interest, as determined by their antigen expression, are assessed for the chromosome or chromosomal region being assessed. The testing is performed on many thousands of whole cells in suspension and analyzed using the Amnis ImageStream^X MarkII (ISXmkII). Both images are generated from the ×60 magnification and digital cameras, and quantitative data are used to assess chromosomal data.

Natasha S. Barteneva and Ivan A. Vorobjev (eds.), *Spectral and Imaging Cytometry: Methods and Protocols*,
Methods in Molecular Biology, vol. 2635, https://doi.org/10.1007/978-1-0716-3020-4_9,

The method was developed on CLL, the most common leukemia in the Western world. CLL has a characteristic phenotype determined by standard multiparameter flow cytometry. The cells are of B-cell lineage and characteristically co-express CD5 and CD23 antigens and show light chain restriction. CLL is also characterized by heterogeneous genetic instability and the prognosis is largely defined by the presence of cytogenetic abnormalities [1]. Cytogenetic aberrations are present in more than 80% of cases, the most common being deletions of 11q, 13q or 17p and trisomy 12. Detection of these abnormalities is important as they assist in clinical decision-making and therapeutic choice. The gold standard for detecting these cytogenetic abnormalities is interphase FISH performed on cell smears. Fluorescently labeled DNA probes bind to specific chromosomal regions and the signal detected manually by fluorescent microscopy. Generally, only 200 cells are assessed and the limit of sensitivity is 3% positive cells.

Immuno-flowFISH has been able to accurately identify +12 and del(17p) in CLL. Thousands of CLL cells, identified by their phenotype, are assessed for specific FISH probe. Thousands of cells (not hundreds) are analyzed at a flow rate of 1000–2000 cells per second. The "extended depth of field" capability of the imaging flow cytometer enables FISH probe signals ("spots") to be resolved and localized within the (stained) nucleus of the immunophenotyped cells. In addition to automated digital analysis, imagery allows for manual inspection of each cell. We describe the methodology for the assessment of chromosomes 12 and 17 in CLL specifically to detect the clinically significant +12 and del(17p) abnormalities. This new methodology, by analysing large numbers of cells with precision enabled by immunophenotyping, provides accurate analysis of these chromosomes and has potential to detect important sub-clones of disease. It can be applied at diagnosis for disease stratification, and following treatment to assess residual disease. These applications will assist clinicians in optimizing therapeutic decision-making and thereby improve patient outcome. The new cytogenomic method described opens other opportunities where the combination of cell phenotype and genotype will assist in clinical diagnostics and in research applications.

2   Materials

2.1   Buffers

1. 10× PBS stock solution. Weigh 160 g sodium chloride (NaCl), 4 g potassium chloride (KCl), 28.8 g di-sodium hydrogen orthophosphate (Na_2HPO_4), and 4.8 g potassium dihydrogen orthophosphate (KH_2PO_4). Dissolve solutes in 1.6 L MilliQ water over low heat on a magnetic stirrer and adjust pH to 7.4. Make up to a final volume of 2 L with MilliQ water and filter

into an autoclaved bottle using a Nalgene bottle top filter then autoclave. Store at room temperature (RT).

2. 1× PBS working solution. Dilute 10× PBS stock 1:10 with MilliQ water, check pH after dilution and adjust to 7.4 if required.

3. Wash buffer: 1× PBS/2% FBS (fetal bovine serum). Defrost FBS (stored at −20 °C) in water bath (from 1 mL aliquots). Add 5 mL 10× PBS and 1 mL FBS to 44 mL MilliQ water.

4. 10% Tween20 solution. Dilute 100 μL of 100% Tween20 in 900 μL of 1× PBS.

5. 1 M Tris-HCl solution. Dissolve 6.075 g of Tris-HCl in 35 mL of MilliQ water and pH to 7.4; make up to 50 mL in MilliQ water.

6. 150 mM NaCl solution. Dissolve 0.4383 g of NaCl in 50 mL MilliQ water.

7. MilliQ water (MQW) or deionized water.

2.2 Cell Preparation

1. Whole blood Ethylenediaminetetraacetic acid (EDTA) collection tube (vacutainer).

2. BD PharmLyse solution (1:10).

2.3 Immuno-flowFISH Detection Panels for Chronic Lymphocytic Leukemia (CLL)

1. *Trisomy 12 (+12)*: Panel 1 – Biolegend CD3-AF647 (clone SK7, Australian Biosearch, Sydney, Australia), BD Horizon CD5-BB515 (clone UCHT2, BD Biosciences, Sydney, Australia), BD OptiBuild CD19-BV480 (clone SJ25C1, BD Biosciences), Vysis CEP12-SpectrumOrange (Abbott Molecular, Sydney, Australia) and SYTOX AADvanced DNA stain (Thermo Fisher Scientific, Sydney, Australia) (Table 1). Panel 2 – BD Horizon CD3-BV605 (clone SK7, BD Biosciences), Biolegend CD5-AF647 (clone UCHT2, Australian Biosearch), BD OptiBuild CD19-BV480 (clone SJ25C1, BD Biosciences), Vysis CEP12-SpectrumGreen (Abbott Molecular) and SYTOX AADvanced DNA stain (Thermo Fisher Scientific) (Table 1).

2. *del(17p)*: BD Horizon CD3-BV605 (clone SK7, BD Biosciences), Biolegend CD5-AF647 (clone UCHT2, Australian Biosearch), BD OptiBuild CD19-BV480 (clone SJ25C1, BD Biosciences), Vysis CEP17-SpectrumGreen (Abbott Molecular), SureFISH 17p12-OrangeRed (Agilent Technologies Dako, Sydney, Australia), and SYTOX AADvanced DNA stain (Thermo Fisher Scientific) (Table 1).

2.4 Immuno-phenotyping, Cell Fixation and Permeabilization

1. Immunophenotyping master mix: +12 Panel 1 25 μL CD3-AF647, 25 μL CD5-BB515, and 25 μL CD19-BV480; +12 Panel 2 and del(17p) Panel 25 μL CD3-BV605, 25 μL CD5-AF647, and 25 μL CD19-BV480.

Table 1
Immuno-flowFISH panels for CLL detection of trisomy 12 and del(17p) on AMNIS ImageStream[X] Mark II

Excitation laser (nM)	AMNIS ISX MKII channel[a]	Emission wavelength (nM)	Fluorophore	CLL biomarker		
				Trisomy 12 panel 1	Trisomy 12 panel 2	Del(17p) panel
N/A	Ch01	BF	N/A	Cell morphology	Cell morphology	Cell morphology
488	Ch02	480–560	BB515, SG	CD5-BB515[b]	CEP12-SG	CEP17-SG
488/561/ 592	Ch03	560–595	SO, OR	CEP12-SO		17p12-OR
488/561/ 592	Ch05	640–745	SYTOX AAD	Nuclear DNA	Nuclear DNA	Nuclear DNA
405	Ch07	430–505	BV480	CD19[b]	CD19[b]	CD19[b]
405	Ch10	595–660	BV605		CD3[b]	CD3[b]
642	Ch11	640–745	AF647	CD3[b]	CD5[b]	CD5[b]

AF647 Alexa Fluor 647 fluorophore, *BF* Brightfield, *BB515* Brilliant Blue 515 fluorophore, *BV480* Brilliant Violet 480 fluorophore, *BV605* Brilliant Violet 605 fluorophore, *CEP* Chromosome enumeration probe, *Ch* Channel, *Del* Deletion, *SG* SpectrumGreen, *SO* SpectrumOrange, *OR* OrangeRed
[a]Only channels relevant to the fluorophores tested are listed
[b]CD3 clone SK7, CD5 clone UCTHC2, CD19 clone SJ25C1

2. Isotype master mix: +12 Panel 1 25 μL AF647, 25 μL BB515, and 25 μL BV480 isotype controls; +12 Panel 2 and del(17p) 25 μL BV605, 25 μL BV605, and 25 μL BV480 isotype controls.

3. BS3 cross-linking solution: 1 mM bis(sulfosuccinimidyl)-suberate (BS3) (Thermo Fisher Scientific) cross-linking working solution (x4 samples/200 μL each). Dissolve 4 mg of BS3 in 350 μL of 1.25× PBS (25 mM PBS) to make 20 mM BS3 stock solution. Store 200 μL aliquots at −20 °C. Dilute 40 μL 20 mM BS3 stock solution in 760 μL of 1.25xPBS (25 mM PBS) to make 1 mM working solution.

4. Quench buffer: 100 mM Tris-HCl pH 7.4/150 mM NaCl solution. Dilute 5 mL of 1 M Tris-HCL with 45 mL of 150 mM NaCl.

5. Fixation and permeabilization buffer: 4% formaldehyde with 0.1% Tween20 in PBS solution (1 mL). Dilute 250 mL of 16% commercial stock formaldehyde solution in 740 μL of 1× PBS, add 10 μL of 10% Tween20, and mix. This buffer should be made at RT and combined for 10 min prior to use. Discard unused buffer.

2.5 Acid Denaturation and FISH

1. Acid solution (0.5 M HCl) – 5 mL, adjust to pH 0.7. Slowly dilute 209 μL of 37%/SG1.18 hydrochloric acid in 1 mL MilliQ water, add MilliQ water to make 5 mL.

2. CEP12 FISH probe master mix: Vysis CEP12-SpectrumOrange or SpectrumGreen probe and Vysis CEP hybridization buffer. For each sample aliquot 7 μL Vysis CEP hybridization buffer, add 2 μL MilliQ water and 1 μL Vysis CEP12 probe.

3. del(17p) FISH probe master mix: Vysis CEP17-SpectrumGreen probe (CEP17SG) and SureFISH 17p-OrangeRed (17pOR) locus-specific probe (LSI) in Vysis LSI hybridization buffer. For each sample aliquot 7 μL Vysis LSI hybridization buffer, add 1 μL MilliQ water, 1 μL Vysis CEP17 probe +1 μL SureFISH 17p probe.

4. 20× SSC stock solution: 3 M NaCl, 0.3 M Na Citrate. Weigh 8.766 g of sodium chloride and 4.9 g sodium citrate and add 50 mL MilliQ water. Adjust to pH 7.0.

5. 2× SSC. Dilute 20× SSC 1:10 in MilliQ water.

6. 0.4× SSC. Dilute 20x SSC 1:50 in MilliQ water. Prewarm to 55 °C for use.

7. Stringency buffer I: 0.1% IGEPAL-CA630 in 2× SSC – 10 mL. Dilute 10 μL of 10% IGEPAL-CA630 in 9.99 mL 2× SSC.

8. Stringency buffer II:0.3% IGEPAL-CA630 in 0.4× SSC – 10 mL. Dilute 30 μL of 10% IGEPAL-CA630 in 9.97 mL 0.4× SSC.

2.6 Additional Materials

1. SYTOX AADvanced nuclear stain (1 μM/1:1000 in PBS stock) (0.2 μM/1:5000 in PBS per test).

2. SYTOX AADvanced 1:1000 working stock (1 μM). Dilute 1 μL SYTOX AADvanced factory stock in 999 μL 1× PBS.

3. SYTOX AADvanced 1:5000 staining stock (0.2 μM). Dilute 20 μL SYTOX AADvanced factory stock in 80 μL 1× PBS.

4. 0.2 mL PCR tube (lo-bind).

5. 1.5 mL clear view microfuge tubes (lo-bind).

6. 5 mL Eppendorf tubes (lo-bind).

7. SPHERO Rainbow Calibration Particles (8 peaks), 3.0–3.4 μm (BD Biosciences).

8. Simply Cellular Anti-Mouse Compensation Standard Controls (Bangs Laboratories Inc., Indiana, USA).

2.7 Instrumentation	1. Eppendorf 5702 centrifuge.

2. Eppendorf 5424 microfuge.

3. Countess I Automated Cell counter (ThermoFisher Scientific) or hemocytometer equivalent.

4. Applied Biosystems Veriti 96 Well Thermal Cycler @ 78 °C for 5 min and 37 °C overnight (24 h), set reaction volume to 10–30 µL.

5. Amnis ImageStream^X MarkII with 405 nm, 488 nm, 561 or 592 nm, 642 nm lasers, 60× magnification, and extended depth of field (EDF).

3 Methods

The following protocol will split across 2 days due to the time required for hybridization.

Day 1

3.1 Preparation of Leucocytes

1. Collect 9 mL peripheral blood into EDTA vacutainer and gently invert eight to ten times.

2. Add 1 mL of whole blood to 10 mL of 1× BD Pharmlyse (*see* **Note 1**). Gently vortex or invert each tube immediately after addition of lysis solution. Incubate at RT for 10 min.

3. Centrifuge at 200 × *g* for 5 min and remove supernatant (*see* **Note 2**).

4. Following red cell lysis, wash cell pellet with 5 mL 1× PBS and centrifuge at 200 × *g* for 5 min, and then remove supernatant. Repeat wash.

5. Resuspend in 250 µL–2 mL PBS (combine cell pellets) for cell counting (*see* **Note 3**).

6. Count cells using a Countess or hemocytometer (use neat or 1: 10 dilution in PBS) with trypan blue, calculate volume to add per test (2–5 × 10^6 cells) based on calculated cell concentration (cells/mL).

3.2 Immuno-phenotyping, Cell Fixation, and Permeabilization

1. Aliquot a minimum of 5 × 10^6 cells (max 1 × 10^7 cells) for each sample in a 5 mL microfuge tube and further resuspend in 600 µL of wash buffer.

2. Centrifuge at 900 × *g* for 3 min and remove supernatant.

3. Add 75 µL immunophenotyping master mix or isotype master mix to cell pellet, resuspend thoroughly and incubate for 30 min at 4 °C protected from light.

4. Add 800 µL wash buffer and centrifuge at 900 × *g* for 3 min then remove supernatant.

5. Resuspend completely in 200 μL of 1 mM BS3 and incubate for 30 min at 4 °C.

6. Add 1 mL of quench buffer and incubate 20 min at 4 °C (*see* **Note 4**).

7. Add 800 μL wash buffer and centrifuge at 900 × *g* for 3 min then remove supernatant. Do not aspirate wash buffer or resuspend cells at this step.

8. Fix samples by adding 250 μL of fix/perm buffer to a loosened cell pellet. Resuspend thoroughly with aspiration and incubate for 10 min at RT.

9. Add 800 μL wash buffer and centrifuge at 900 × *g* for 3 min then remove supernatant.

10. Resuspend in 800 μL of wash buffer (*see* **Note 5**).

3.3 Acid Denaturation and Probe Hybridization

1. Centrifuge at 900 × *g* for 3 min and remove supernatant.

2. Slowly add 100 μL of acid solution and mix gently, incubate for 20 min at RT.

3. Add 3 mL ice-cold PBS and centrifuge at 600 × *g* for 10 min then remove supernatant.

4. Add 1 mL of wash buffer, washing down the side of the tubes, centrifuge at 900 × *g* for 3 min, and remove supernatant.

5. Resuspend in 150 μL of stringency buffer I. Transfer cells to 0.2 mL lo-bind PCR tube and centrifuge at 950 × *g* for 3 min.

6. During the previous step, prewarm CEP12 or del(17p) FISH probe master mix in a separate PCR tube at 37 °C for 10 min.

7. Remove all excess stringency buffer I from samples in PCR tubes (*see* **Note 6**).

8. Resuspend cells in the pre-warmed FISH probe master mix and thoroughly resuspend cells.

9. Place samples in thermal cycler, set reaction volume setting to 10–30 μL to ensure even heating, denature at 73 °C for 5 min then hybridize the FISH probe at 37 °C for at least 16 h for CEP12 analysis, **OR**, at 78 °C for 5 min and then hybridize at 42 °C for at least 24 h for del(17p) analysis (*see* **Note 7**).

Day 2

3.4 Post-hybridization Wash and DNA Stain

1. Add 150 μL of stringency buffer I, mix gently, and transfer to 1.5 mL lo-bind microfuge tubes. Centrifuge at 950 × *g* for 3 min and remove supernatant.

2. Resuspend in 200 μL of stringency buffer II (prewarmed to 55 °C) and incubate for 5 min at 55 °C to degrade excess probe.

3. Add 800 μL wash buffer and centrifuge at 900 × *g* for 3 min then remove supernatant.

4. Resuspend in 30 μL SYTOX AADvanced DNA stain (1:5000 working stock) and incubate for 20 min at RT.

3.5 AMNIS ISX MarkII Instrument Setup

1. Allow an ASSIST calibrated AMNIS ISX MarkII (with INSPIRE 4.2 software) to stabilize based on graphical tracking of frequency versus Brightfield Root Mean Square (RMS) histogram (Fig. 1a). A core stability value of at least 60 for in-focus beads is desired with an instrument auto-focus of a range of (+/−) 2 for core depth in the z-plane. Adjust core position (x-plane) alignment if required. Ensure that Brightfield "set-intensity" setting has been automatically adjusted for Channels 1 and 9 (Fig. 1b).

2. Set the following acquisition parameter values (Fig. 1b and Table 1):

    (a) Magnification: 60× objective with extended depth of field (EDF) up to in-focus range of 16 μM. Set fluidics to "lo" speed and "hi" sensitivity with core size of 6 μM.

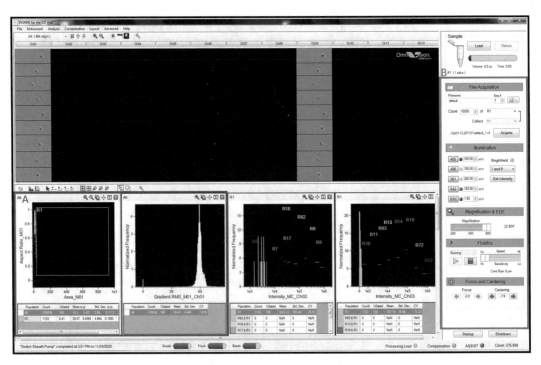

Fig. 1 AMNIS INSPIRE version 4.2 interface for immuno-flowFISH data acquisition setup. (**a**) Identification of events (e.g., beads) based on Brightfield Area versus Aspect Ratio and tracking of core focus based on brightfield Gradient Root Mean Square (RMS) histogram (red box). A core stability value of 60 or higher for in-focus beads is desired for stabilization of instrument prior to data acquisition. (**b**) File acquisition, illumination, magnification and image capture parameters for optimal CLL immuno-flowFISH analysis (green box)

(b) Ensure that all channels that are required for the capture of fluorescence signals (i.e., antibody and probe) for the experiment are checked/active. For CLL immuno-flowFISH panels this will include Channels (Ch) 01, 02, 03, 05, 06, 07, 10, and 11 (*see* **Note 8**). See below for excitation and emission spectra for the fluorescent markers of interest.

(c) Illumination (Brightfield): ON for Channel 1 and/or 9 (OFF for compensation controls).

(d) Illumination (SSC): 1.5 mW 785 nm laser to provide a scatter signal (OFF for compensation controls).

(e) Illumination (405 nm): 100 mW 405 nm laser to excite BD Horizon BV480 and capture emission in the wavelength range 430–505 nm (Ch07); in addition, to excite BD Horizon BV605 and capture emission in the range 595–660 nm (Ch10).

(f) Illumination (488 nM): 100 mW 488 nm laser to excite BD Horizon BB515 or SpectrumGreen and capture emission in the range 480–560 nm (Ch02); in addition, to excite SYTOX AADvanced DNA stain and capture emission in the range 660–740 nm (Ch05).

(g) Illumination (561 or 592 nm): 200 mW 561 nm or 592 nm laser to excite Vysis SpectrumOrange conjugated chromosome 12 enumeration probe or SureFISH del (17p)-OrangeRed probe, and capture emission in the range 560–595 nm (Ch03).

(h) Illumination (642 nM): 120 mW 642 nm laser to excite Biolegend AF647 and capture emission in the wavelength range 640–745 nm (Ch11).

3. Load SPHERO Rainbow Calibration Particles prepared with manufacturer's instructions for longitudinal tracking of daily instrument performance (*see* **Note 9**).

(a) Acquire 1000 events on gate set on Area versus Aspect Ratio (geometric mean of around x = 48, y = 0.8) gate beads (Fig. 2a) and display in fluorescence intensity histograms to establish 6–8 peak histogram profiles (i.e., frequency versus fluorescence intensity) in the relevant channels used in CLL immuno-flowFISH panels (i.e., Ch01, 02, 03, 05, 06, 07, 10, 11). Ensure the fluorescence intensity values of 6–8 peaks per channel are matched with reference ranges based on previous instrument benchmarks (example illustrated in Fig. 2b).

(b) If required, adjust laser powers for excitation utilized in the experiment to produce 6–8 peak coefficients/values comparable to previous SPHERO Rainbow Calibration

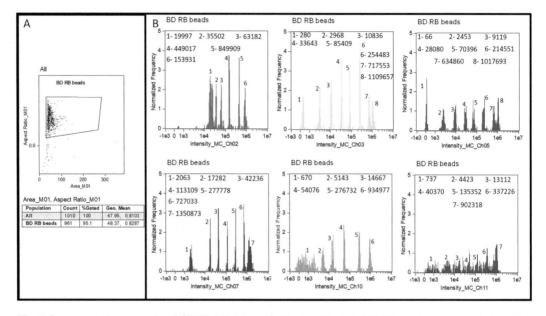

Fig. 2 Representative example of SPHERO Rainbow Calibration Particle (BD RB) performance data from the AMNIS ISXmkII at the University of Western Australia. (**a**) Detection of BD RB events based on brightfield Area versus Aspect Ratio gating to establish 6–8 peak histogram profiles. (**b**) BD RB frequency versus fluorescence intensity 6–8 peak histogram profiles and peak fluorescence intensity values/coefficients for detection channels (Ch) Ch02, 03, 05, 07, 10, and 11 which are used in CLL immuno-flowFISH

Particle performance data or representative UWA examples of ISXMarkII values provided (*see* **Note 10**; Fig. 2b).

3.6 Sample Acquisition

1. Ensure that AMNIS ISXMarkII is allowed to stabilize as per criteria above following SPHERO Rainbow Calibration Particle calibration and then after loading of CLL samples. Record from 10,000 to 200,000 events from gate set for singular cells (<105:>0.8 Area vs Aspect Ratio; to the right of the Speed-Bead population). On average, each sample can be acquired at a rate of 1–300 cells per second (i.e., 1–10 min/10,000 event collection for each sample) (*see* **Note 11**; Fig. 1a).

2. Following CLL sample data acquisition of, acquire data for unstained, isotype, and single-stained compensation controls with Simply Cellular Compensation Control particles. Compensation controls may include FITC and PE (for Vysis and SureFISH probes), BD Horizon BV480/BB515/BV605, Biolegend AF647 (for cell surface markers), and SYTOX AADvanced-stained cells (for nuclear marker) using identical laser settings in the absence of Brightfield and 785 nm laser illumination, and calculating a compensation matrix for spectral compensation using IDEAS v6.0 software (Amnis, Luminex Corporation, Seattle, USA). A reliable compensation matrix will require acquisition of at least 500 "positively

stained" beads and this can be done directly with gating on the positive peak of "All" events in the channel of interest (e.g., CD19-BV480 positively stained beads in channel 7 (430–505 nm)).

3.7 IDEAS Data Analysis

1. Perform immuno-flowFISH CLL data analysis using IDEAS image analysis software v6.0 using compensated data [2, 3].

2. Select focused images using the Gradient RMS feature which measures the sharpness quality of an image by detecting large changes in pixel values in the image where cells with better focus have higher Gradient RMS values. Gate in-focus events based on a value of around 60 and above (*see* **Note 12**; Fig. 3a).

3. From the in-focus events, identify single cells in a scatter plot of the Aspect Ratio versus Brightfield Area (Fig. 3b). The Aspect Ratio features the Minor Axis of the object or cell divided by the Major Axis and describes how circular or oblong an object is. Single nucleated cells have a high Aspect Ratio and low Area value (e.g., around x = 105, y = 0.8), which permits the exclusion of cell doublets, debris, and clumps (*see* **Note 13**).

4. Identify nucleated non-dividing cells from in-focus, single cells using a SYTOX AADvanced fluorescence intensity histogram by excluding cells with high fluorescence intensity (dividing cells, dead cells and cell clumps) (Fig. 4a).

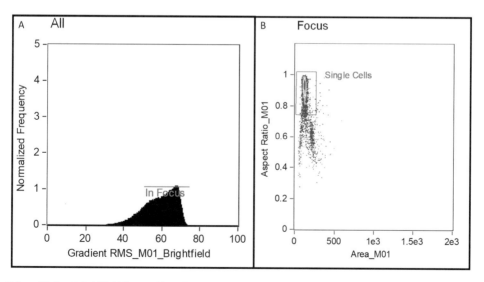

Fig. 3 Quantitative brightfield image identification of in-focus and single cells. (**a**) Histogram of Gradient RMS feature enables detection of in-focus images of events where higher Gradient RMS values (>60) equate to sharper, better quality images of cells. (**b**) Single cells can be identified using a scatter plot of the Aspect Ratio versus Brightfield Area, where single cells have a high Aspect Ratio and low Area value (e.g., around x = 105, y = 0.8), which permits the exclusion of cell doublets, debris and clumps

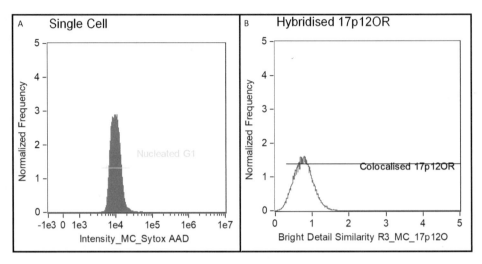

Fig. 4 Quantitative nuclear DNA stain assessment identifies nucleated cells with specific nuclear co-localized FISH signals. (**a**) Analysis of fluorescence intensity histogram of SYTOX AADvanced staining identifies nucleated non-dividing cells by excluding cells with high fluorescence intensity (i.e., dividing cells, dead cells, and cell clumps). (**b**) Histogram of Similarity feature calculation with SureFISH 17p12-Orange-Red (Ch03) signals with the SYTOX AADvanced fluorescence masks enables identification of events with specific FISH probe signals within the nucleus (>0.65). Non-specific probe hybridized to DNA from disrupted cells adhered to the cell membrane have a low similarity score and can be excluded

5. Determine co-localization of FISH probe signals for Vysis CEP12/CEP17 SpectrumGreen (Ch02), Vysis CEP12-SpectrumOrange or SureFISH 17p12-OrangeRed (Ch03) with the nuclear stain (SYTOX AADvanced fluorescence) using the Similarity feature calculation. This feature measures the degree to which two images (FISH probe signal and nuclear stain fluorescent pixels) are linearly correlated within a masked region. Gated events with FISH probe signals within the nucleus, will have a high similarity score (>0.65). In contrast, non-specific probe hybridization events such as probe or DNA adhering to the cell membrane (e.g., disrupted cells) will result in a lower similarity score and can be excluded (*see* **Note 14**; Fig. 4b).

6. Utilize the Spot Count feature to count the number of CEP12/CEP17-SpectrumGreen (Ch02), Vysis CEP12-SpectrumOrange or SureFISH 17p12-OrangeRed (Ch03) spots per cell using a mask that incorporates the "Spot" and "Intensity" image analysis functions calibrated for each probe in the experiment. The "Spot" and "Intensity" masks can be generated in the mask manager for each probe. Define the mask function with the following parameters (Figs. 5 and 6):

   (a) Function: Spot, Mask: Bright (M02 and Ch02 for SpectrumGreen and M03 and Ch03 for SpectrumOrange or OrangeRed).

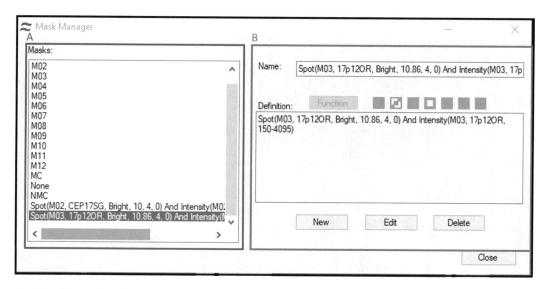

Fig. 5 Setting up the "Spot" and "Intensity" image analysis functions for the Spot Count feature to enumerate FISH "spots" per cell. (**a**) The "Spot" and "Intensity" masks can be generated in the mask manager for each probe (red box). (**b**) Navigating the mask manager to define the mask function (green box)

(b) Select a representative object example of an in-focus, single cell with specific FISH signals as a reference image for mask calibration.

(c) Set minimum radius of "0" and a maximum radius of "4" (taking into account the diameter of the FISH spots, 0–4 have been observed for specific Vysis and SureFISH FISH signals), then adjust "Spot to Cell Background Ratio" from a range of 1.00–30.00 (increasing ratio). This ratio is the spot pixel value divided by the mean background value in the bright detail image. For this adjustment, pay attention to the software-defined blue masked area as you adjust (Fig. 6a). The mask will get smaller as the ratio increases, the final value should produce a mask that precisely or tightly overlays with the observed fluorescent spots in the probe channel (Fig. 6b). Press "OK" to define this component of mask function, then append "And" operator to definition to select and incorporate the "intensity" component to the function. After "intensity" is selected as function, repeat A and B selections above, then set "maximum" intensity to 4095 (or max value) and adjust or increase the "minimum" intensity, paying attention to the software-defined blue masked area, which will get smaller as the minimum intensity is increased (Fig. 6c). This can be assisted by the software by "pointing the cursor" over individual pixels of the probe/cell image to determine specific FISH pixel intensity values to calibrate the

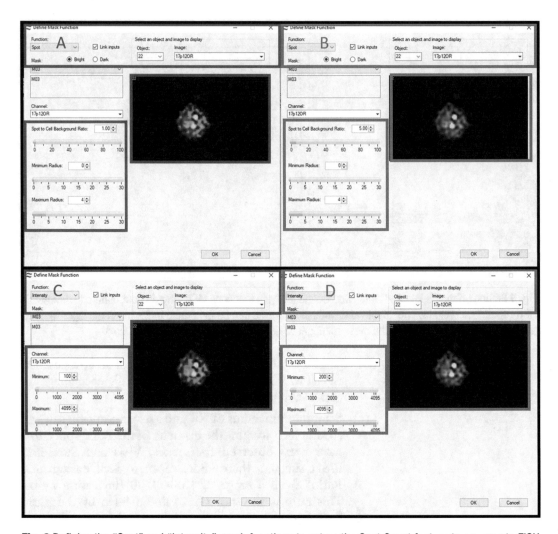

Fig. 6 Defining the "Spot" and "Intensity" mask functions to set up the Spot Count feature to enumerate FISH "spots" per cell. (**a**) Suboptimal definition for the "Spot" image analysis mask (red box). (**b**) Optimal definition for the "Spot" image analysis mask (green box). (**c**) Suboptimal definition for the "Intensity" image analysis mask (red box). (**d**) Optimal definition for the "Spot" image analysis mask (green box)

intensity limit. The final value should produce a mask that precisely or tightly overlays with the observed fluorescent spots in the probe channel (from 170–600) (Fig. 6d). Press "OK" to complete function and confirm overall definition for software masks to identify FISH signals (*see* **Note 15**).

7. Open a new "feature" in the feature manager, select a "single" "spot" feature type (rename if necessary) and then select spot and intensity function masks generated above and check for "Four" (Fig. 7). The Spot Count Feature algorithm examines the connectivity of each pixel based on whether it is connected

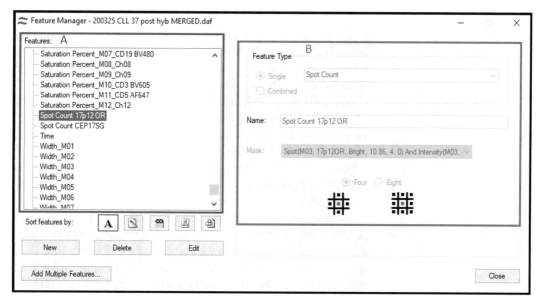

Fig. 7 Setting up the Spot Count feature to enumerate FISH "spots" per cell with masks that incorporates the "Spot" and "Intensity" image analysis functions calibrated for each probe in an immuno-flowFISH experiment. (**a**) The Spot Count feature calculation can be generated in the feature manager for each probe (red box). (**b**) Selecting the feature type and mask functions to program the feature calculation (green box)

to a particular spot or the background. Click "OK" and graph a new "histogram" on in-focus, single, and probe co-localized cells for normalized frequency versus spot count feature defined above. Gate-generated histograms for 0, 1, 2, 3, and 3+ increments on the x-axis to enumerate percentage of cells in each category. Current references range for immuno-flowFISH testing with Vysis CEP12, CEP17, and SureFISH 17p12 probes for cells without numerical or structural chromosomal number changes (e.g., healthy cells or T lymphocytes): 0-spots = < 1%, 1-spot = around 10–20% (majority overlapping), 2-spots = 70–80%, 3-spots = 1–5%. If the del(17p) panel is used then the Vysis CEP17-SpectrumGreen spot count will provide an internal control to the SureFISH 17p12-OrangeRed spot counts (*see* **Note 16**).

8. Further quantitative verification is required for software-generated spot counts by single parameter histograms comparing the measured fluorescence intensity of FISH signals for each of the spot count populations (i.e., 1-spot versus 2-spot versus 3-spot). Changes in measured fluorescence intensity correspond to the number of hybridized probes, allowing correction of overlapping signals (i.e., 2-spots that appear as 1-spot) inherent in 2-dimensional image projections to determine the true number of specific FISH spots and reliable distinction of monosomy and disomy subpopulations [4].

9. Identify CLL cells, B lymphocytes, and T lymphocytes from in-focus, single, and probe co-localized cells for immuno-flowFISH analysis. This is achieved through gating populations based on the fluorescence intensity of CD19-BV480/BB515 (normal B lymphocytes and CLL cells), CD5-BB515/AF647 (CLL cells and T lymphocytes), and CD3-BV605/AF647 (T lymphocytes). Graphing CD19 versus CD3 and CD19 versus CD5 bicolour scatterplots will enable multi-parameter distinction of CLL cells (CD19+/CD5+, CD3-), normal B lymphocytes (CD19+, CD3-/CD5-) and T lymphocytes (CD19-, CD3+/CD5+). Apply the custom Spot Count feature generated above for each gated cell population of interest to specifically compare the cytogenetic status of the CLL cells (CD19+/CD5+) with the normal lymphocytes to determine the status of chromosomes 12 and 17 for possible trisomy 12 or del(17p) (*see* **Note 17**). A representative example of a CLL with del(17p) is shown (Fig. 8) which highlights that precision FISH analysis of cells by immunophenotype is important. The CLL population (CD19+/CD5+) has an aberrant del(17p) cytogenetic status (majority 1-spot for 17p12) (Fig. 8b, d) compared to the normal T lymphocyte population (CD19-, CD3+/CD5+) which have 2 copies of 17p (majority 2-spot for 17p12) (Fig. 8a, c).

10. We recommend using image galleries (Fig. 9) to view all features as an ongoing reference to refine gating and analysis strategies (i.e., does the observed visual imagery correlate with the software-driven quantitative analyses or definitions?). Digital imagery enables direct visualization and verification of the multi-parametric immuno-flowFISH features used in CLL assessment at the single-cell resolution. Both single parameter and composite image galleries for incorporating each channel/biomarker (i.e., brightfield morphology, immunophenotype, SYTOX DNA stain, and probe FISH signals) should be inspected in context with each other together to confirm cell biology, type, and chromosomal status. This can be used for a final "check" for cell autofluorescence, disrupted cell morphology or viability, non-specific staining of antibody and probe that would otherwise compromise accuracy and precision of the test. For example: normal B and T lymphocytes will have 2 chromosome 12 enumeration probe signals (2 CEP12 spots, disomy 12); CD19+/CD5+ CLL cells with trisomy 12 will have 3 spots for CEP12; and CD19+/CD5+ CLL cells with del(17p) will have 2 CEP17 signals and a single co-localized chromosome 17p12 signal (2 green spots and 1 yellow spot). Merge of the immunophenotype, probes, and SYTOX AAD-vanced nuclear marker images (overlay) shows co-location of FISH spots within the nucleus confirming specificity and cell

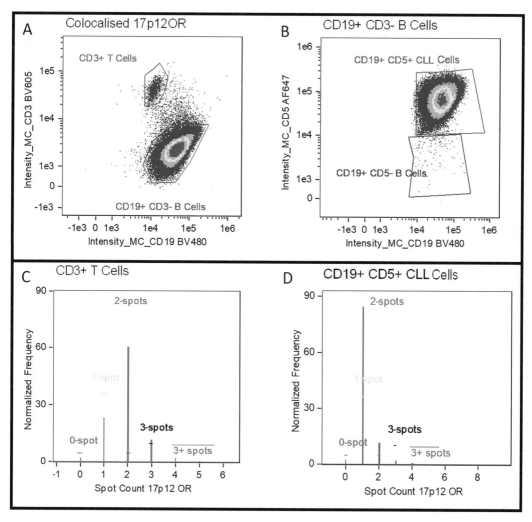

Fig. 8 Immuno-flowFISH gating strategy to identify del(17p) in CLL. (**a**) Bi-color dot plots are used to identify T lymphocytes, B lymphocytes and CLL cells based on CD19 versus CD3 expression and fluorescence intensity. (**b**) Bi-color dot plots shows CD5 expression discriminates CLL cells from normal CD5- B lymphocytes. (**c**) Application and enumeration of the custom Spot Count feature shows the majority of CD3+/CD5+ T lymphocytes to have a 2-spot (disomy) count, whereas in (**d**) the majority of CLL cells (CD19+/CD5+) have 1-spot count indicating (del(17p))

type of interest. Save overall IDEAS analysis templates and apply to all future sample analyses to standardize workflow and data reproducibility to generate reliable interpretation of CLL chromosomal data (i.e., generate reference ranges and appropriate in-house laboratory cut-off values).

11. To standardize immuno-flowFISH detection of chromosome 12 or 17 in CLL samples based on single-cell characteristics and population size, "Spot Count Ratios" can be calculated from the spot count statistics for each subpopulation [3]. This

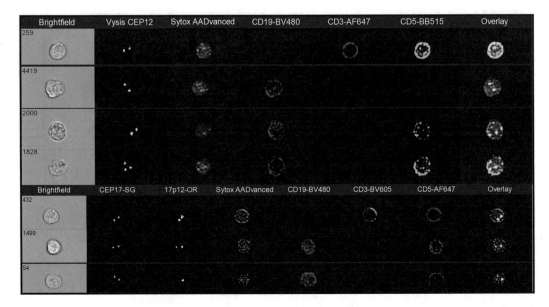

Fig. 9 Representative image galleries of normal and CLL cells analyzed by immuno-flowFISH illustrating different patterns. The top 4 galleries show FISH staining patterns for CEP12 chromosome 12 enumeration probe. Cell 259 is a CD3+/CD5+ T lymphocyte and cell 4419 a CD19+ B lymphocyte, both with the normal 2 FISH spots. Cell 2000 is a CLL cell (CD19+/CD5+) with 2 FISH signals for CEP12 and cell 1828 a CLL cell with trisomy 12 shown by the 3 FISH signals. The lower four galleries show CEP17 and 17p12 probes. Cell 432 is a normal CD3+/CD5+ T lymphocyte with 2 CEP17 signals and 2 co-localized chromosome 17p12 signals. Cells 1499, 54 and 2529 are all CLL cells (CD19+/CD5+). Cell 1499 has normal 2 normal CEP17 signals and 2 co-localized 17p12 signals. Cell 54 is a CLL cell with 1 spot for 17p12 overlying CEP17, indicating loss of the 17p region from one of the chromosomes indicating del(17p). The presence of 2 spots for CEP17 indicates 2 copies of the chromosome. Merge of the immunophenotype, probes and Sytox AADvanced nuclear marker images (overlay) shows co-location of FISH spots within the nucleus confirming specificity

is achieved by calculating mean spot counts per cell in each immunophenotyped population (using the percentage or absolute number of cells with $0/1/2/3/3+$ spots) where the total number of spots is normalized to phenotypic subpopulation size:

$$\frac{\text{CLL}(\text{CD}19+/\text{CD}5+, \text{CD}3-)}{\text{Tlymphocytes}(\text{CD}19-, \text{CD}3+/\text{CD}5+)}$$

Specifically, the immuno-flowFISH calculation and criteria for cytogenetic assessment and normalization is as follows:

(a) Spot count ratio = Mean spot count CLL cells divided by the mean spot count of normal T lymphocytes.

(b) Normal cells will have a mean spot count of close to two for both chromosomes 12 and, 17 and the 17p region. CLL cells with no alteration in chromosome 12 or 17 ploidy, and with retained 17p region will also have 2 FISH signals and hence a ratio of 1 when compared to the T lymphocyte population.

(c) CLL cells with trisomy 12 (+12) will have a mean spot count of between 2 and 3, due to the additional signal. The spot count ratio when compared with T lymphocytes will therefore be between 1 and 1.5.

(d) CLL cells with loss of material from the short arm of chromosome 17 or del(17p) will have a mean 17p12 spot count of less than 2. The CEP17 mean spot count will be normal at around 2 due to preservation of the centromeric region; the CEP17 probes therefore functions as an internal control for chromosome 17. The spot count ratio, comparing 17p signals with CEP17 in the CLL cells or the T lymphocytes will therefore be less than 1 (due to the reduction in 17p FISH spots). In contrast, the spot count ratio of CEP17 when compared to the T lymphocyte population will be normal at 1.

Representative image galleries for all detected cells in CLL samples based on UWA immuno-flowFISH testing are shown (Fig. 9).

4 Notes

1. For samples with a low leucocyte count or if you require multiple samples for assessment the processing volume can be adjusted to 5 mL of blood and 45 mL of 1× BD PharmLyse. HAZARD: Use filtered pipette tips and appropriate personal protective equipment when pipetting blood samples or removing supernatant until samples have been fixed with formaldehyde.

2. Centrifuge at 200 × g for 10 min if processing 5 mL of blood in 45 mL of 1× BD PharmLyse and then remove supernatant. Resuspend pellet in 5 mL of 1× PBS and transfer into 15 mL tube.

3. May need to adjust volume to 5 mL for patients with high leucocyte count.

4. We have found cells must be handled gently at this stage of the protocol to prevent lysis. Add quench buffer slowly, drop by drop, and do not aspirate. Drops of quench buffer should be added to the side of the tube not directly to the cell pellet.

5. Remove an aliquot of test samples for analysis of "post-stain" immunophenotyping on AMNIS ISXmkII. Isotype control samples are removed for analysis on AMNIS ISXMarkII and do not proceed to acid denaturation and hybridization.

6. It is important to remove all excess 0.1% IGEPAL in 2xSSC buffer prior to adding FISH probe master mix as residual buffer can change the hybridization mix ratio and reduce

hybridization efficiency. Pulse spin samples and use a p10 pipette to remove as much excess buffer as possible.

7. It is essential to ensure that adequate hybridization times are utilized for each probe to ensure good and standardized quality of FISH signals for quantitative detection. All chromosome enumeration probes (CEP) listed for CLL immuno-flowFISH analysis should receive at least 16 h of hybridization time at either 37 °C or 42 °C. Shorter hybridization times are achievable due to the relatively larger size of CEP probes and their potential binding regions (>1000 kb) with kinetics that rapidly generate/amplify signals for detection. Longer hybridization times are required for LSI such as 17p12 as these probes are relatively smaller and hybridize to a smaller, more specific region of the chromosome (<800 kb). Thus, the kinetics of hybridization is relatively slower and hybridization times of at least 24 h at either 37 °C or 42 °C are required to generate good quality locus-specific FISH signals for immuno-flowFISH detection. Working ranges for hybridization times can be optimized for each laboratory's purposes. This can be based on time feasibility and sample/probe input requirements, while still using the UWA lower limit of hybridization times listed above for both CEP and LSI probes. However, we encourage that sample hybridization times be standardized longitudinally (between experiments) so that reproducible FISH results are achieved for immuno-flowFISH analysis and interpretation.

8. To reduce file size and improve file transfer and analysis time of post-acquisition data files, unused channels (e.g., Ch04, 08, 12) can be turned off (unchecked) prior to acquisition.

9. Ensure that the same lot number/factory batch of SPHERO Rainbow Calibration Particles is used to standardize the assessment of instrument performance and calibration.

10. The configuration of all instruments and operating conditions (e.g., laser alignment) can deviate from factory defaults and between laboratories. Therefore, instrument calibration using acquisition standards such as SPHERO Rainbow Calibration Particles should be adapted in-house to optimize the utility and experimental application of your ISXmkII. While we provide a UWA reference set of SPHERO Rainbow Calibration Particle data, calibration templates with appropriately adjusted acquisition settings should be generated to maximize acquisition quality and reproducibility for each laboratory. Always reference instrument calibration results to interpret longitudinal experimental results.

11. It is important to ensure that core stability is conserved during acquisition of samples (i.e., brightfield RMS focus tracking,

core auto-focus, and x-y alignment parameters). Core stability and image capture stream quality may shift due to background turbulence generated by the cell samples. This includes samples with high cellularity, poorer quality, or "stickiness" (high debris and clumps – as seen in Brightfield Area versus Aspect Ratio) or result from long acquisition times (e.g., sample build-up in core). This may benefit from adjustment of acquisition volume (i.e., reduce cell concentration) or "priming" of samples for throughput or manual focusing and centering of core to ensure uniformity of acquisition to maximize quality of data generated. Adjust your acquisition sample volume, numbers of cells to be collected, or collection time to the CLL sample investigation objective. Disease monitoring, residual disease post-therapy/low yield samples, and rare cell detection applications will benefit from longer acquisition times to ensure that sufficient events are collected for more sensitive assessment.

12. Based on the data collected at UWA, cells that are of "good" or "sufficient" focus are essential to IDEAS-driven FISH analysis and reproducible spot-counting. This should include events with at least a 60 Brightfield RMS value. Gating events below this threshold is discouraged as events that are less in focus have FISH signals that are more variable or heterogeneous in quality and are generally "dimmer" (especially for LSI like SureFISH 17p12). This may confound spot-counting results due to the binary nature of the IDEAS masking strategy.

13. The scatterplot analysis of Area versus Aspect Ratio will also highlight differences in the morphological dimensions of cellular events. This may be due to the biology of the cells in the sample or highlight critical changes in protocol alterations (e.g., starting sample quality, preservation, or suboptimal processing). Thus, it may be better to set a less stringent gate (i.e., covering a broader spectrum of clustered events) so that biologically significant cells of interest in the sample are not excluded in downstream analysis. Cell doublets and clumps not excluded at this stage can be removed in the SYTOX AADvanced DNA stain analysis.

14. Issues with FISH can be highlighted by the Similarity feature calculation when co-localization of FISH probe signals with the nuclear stain results in low values (i.e., <0.65) or a graph with a broad spread (i.e., high CV) around the 0.65 threshold. This indicates suboptimal probe hybridization (of specific FISH signals) or irregular cell preparation artifacts such as increased "stickiness" so that high numbers of FISH spots (non-specific) reside outside of the nucleus or on the outer surface of the cell. High non-specific probe staining can result from compromised cell quality, preparation, preservation, and FISH probe integrity.

15. For all subsequent experiments and analyses, the "Spot" and "Intensity" masks should be interrogated (and adjusted if necessary) to reflect precise software-based spot enumeration. The IDEAS masking algorithm and feature calculations are sensitive to subtle variations of FISH hybridization efficiencies and the pre-set analysis templates with previous samples (even the same sample type) may be suboptimal for the calculation of spot counts in samples analyzed on alternate days. This would alter software-based quantification of true-positive, false-positive and false-negative FISH signals. A pre-set FISH analysis template should serve as a universal constant or a reference to previous FISH performance, then altered accordingly to account for subtle changes in FISH signal intensities due to cell preparation (e.g., variable cell number, age, preservation of cell and DNA, variable quality, and chemical alterations), compensation matrix and probes used (e.g., pipetting error, age and reagent lot changes). Factors such as baseline cell autofluorescence and probe "brightness" in the probe channels should be considered when setting signal-to-background ratios and lower intensity cut-off thresholds. For example, SpectrumOrange FISH probes need a higher cut-off in channel 3 than locus-specific SureFISH 17p12-OR probes which are relatively smaller and "dimmer" (i.e., adjust with less stringency in masking strategy).

16. The spot count distributions (0/1/2/3/3+ spots) presented here serve as a reference based on 6 healthy and 60 CLL samples reproducibly assessed at UWA. Large discrepancy in spot-count number compared to this reference range or between experiments for healthy controls may be due to reasons described above and related to acquisition, sample, or processing quality. CLL samples validated with appropriate controls and FISH analysis that deviate from "normal" spot-count ranges are of biological/clinical significance (e.g., higher 1-spots due to deletion in copy number as opposed to consistent baseline level of overlapping spots.

17. Post-stain controls, negative controls, isotype, and healthy sample controls should always be used in each experiment to reliably identify and quantify subpopulations of CLL cells, B lymphocytes, and T lymphocytes by immunophenotype. This template utilizes diagnostic flow cytometry principles that are standardized for the identification of neoplastic cells of interest from internal normal/healthy references cells and permits direct correlation to clinical flow cytometry data (if available). This will assist in reproducible cell detection and the detection of rare cells, determine if any cell populations have been altered or lost, and enable longitudinal tracking of immuno-flowFISH assessment and instrument performance. Expression levels of

CD19 and CD3 on CLL cells, B lymphocytes, and T lymphocytes should remain relatively consistent between experiments and samples, with a fluorescence intensity of at least 1×10^5 with a complete surface "ring" like appearance under immuno-flowFISH visualization. These are robust ubiquitous markers used in all clinically relevant lymphocyte panels. Dead or compromised cells can be excluded based on high staining intensity and autofluorescence in the detection channels capturing CD19 and CD3 expression (high dual positivity). For immuno-flowFISH, the positive expression cut-off is relatively high ($>1 \times 10^4$) compared to standard flow cytometry assays as higher autofluorescence across the detection range is produced by FISH processing. Both X and Y-axis ranges may need to be expanded from default IDEAS settings as some neoplastic cell populations are relatively bright or skewed due to suboptimal compensation, and therefore may be "hidden" from graphical view. The expression of CD5 remains relatively uniform for T lymphocytes however it is known to be heterogeneous in CLL cells. The expression of CD5 in CLL has been identified to range from 1×10^4 to 1×10^6 with a "punctate" pattern by immuno-flowFISH assessment at UWA. For this reason, we discourage CD5 antibody titration for immuno-flowFISH as it may confound analysis for "dim" CD5 expression cases. The immunophenotype of the T lymphocytes, normal B lymphocytes, and the post-stain controls should always be referenced when assessing CD5 expression in CLL cells. We recommend that all populations including the CD19+/CD5- ("B lymphocytes") as well as the CD19-/CD5-/CD3- population (e.g., granulocytes) should be interrogated due to the potential for biological and sample heterogeneity. "Back gating" of spot-count populations (e.g., 3-spot CEP12 or 1-spot del(17p)) to cell immunophenotype will also help decode the clonal diversity present.

References

1. Döhner H, Stilgenbauer S, Benner A, Leupolt E, Kröber A, Bullinger L, Döhner K, Bentz M, Lichter P (2000) Genomic aberrations and survival in chronic lymphocytic leukemia. NEJM 343:1910–1916

2. Hui H, Fuller KA, Chuah H, Liang J, Sidiqi H, Radeski D, Erber WN (2018) Imaging flow cytometry to assess chromosomal abnormalities in chronic lymphocytic leukaemia. Methods 134–135:32–40

3. Hui HYL, Clarke KM, Fuller KA, Stanley J, Chuah HH, Ng TF, Cheah C, McQuillan A, Erber WN (2019) "Immuno-flowFISH" for the assessment of cytogenetic abnormalities in chronic lymphocytic leukemia. Cytometry A 95:521–533

4. Minderman H, Humphrey K, Arcadi JK, Wierzbicki A, Maguire O, Wang ES, Block AW, Sait SNJ, George TC, Wallace PK (2012) Image cytometry based detection of aneuploidy by fluorescence in situ hybridization in suspension. Cytometry A 81:776–784

Chapter 10

Quantifying Golgi Apparatus Fragmentation Using Imaging Flow Cytometry

Inbal Wortzel and Ziv Porat

Abstract

Unlike the common conception of the Golgi apparatus as a static organelle, it is, in fact, a dynamic structure, as well as a sensitive sensor for the cellular status. In response to various stimuli, the intact Golgi structure undergoes fragmentation. This fragmentation can yield either partial fragmentation, resulting in several separated chunks, or complete vesiculation of the organelle. These distinct morphologies form the basis of several methods for the quantification of the Golgi status. In this chapter, we describe our imaging flow cytometry-based method for quantifying changes in the Golgi architecture. This method has all the benefits of imaging flow cytometry—namely, it is rapid, high-throughput, and robust—while affording easy implementation and analysis capabilities.

Key words Golgi, Imaging flow cytometry, Golgi fragmentation, Golgi structure

1 Introduction

Discovered in 1898 by Italian physician Camillo Golgi, the Golgi complex is a key component of the endomembrane system that is absent in prokaryote organisms. Therefore, from an evolutionary perspective, it is a defining feature that separates eukaryotes from prokaryotes [1]. In all eukaryotes, the Golgi consists of cistern-like membranes that connect to form mini-stacks. Whereas in lower eukaryotes, such as protists, plants, and invertebrates, these mini-stacks are scattered throughout the cell, vertebrates have developed a system that connects them into a 'ribbon-like' structure localized in a juxtanuclear position during interphase [2].

Several subcellular organelles (e.g., nucleus, mitochondria, endoplasmic reticulum) are known for their potential to sense various cell processes [3–6], in particular, stress. As has recently been shown [2], the Golgi complex, too, is able to coordinate and sense multiple cellular events, thanks to its central location in the cell and it serves as a hub for almost all endocytic pathways. When

Natasha S. Barteneva and Ivan A. Vorobjev (eds.), *Spectral and Imaging Cytometry: Methods and Protocols*,
Methods in Molecular Biology, vol. 2635, https://doi.org/10.1007/978-1-0716-3020-4_10,
© Springer Science+Business Media, LLC, part of Springer Nature 2023

sensing a change in cellular conditions, the 'ribbon like' structure of the Golgi undergoes reorganization, forming a partially fragmented (similar to lower eukaryotes) or fully fragmented formation (a total dispersion on the organelle). These changes in the Golgi architecture happen in response to physiological processes such as mitosis, apoptosis, and migration [7–9], as well as pathological conditions, including cancer [10, 11], neurological diseases [12–14], stress [15, 16], and infection [17, 18].

As Golgi complex is a sensitive, visible indicator of the cellular status, it is not surprising that multiple methods to quantify Golgi fragmentation have arisen over the years. These methods can be categorized into two groups, according to their application. The first group includes small-scale and/or live-cell imaging. Some of the main methods in this group are conventional immuno-fluorescence (IF) followed by manual counting, fluorescence recovery after photobleaching (FRAP) [19], and laser nano-surgery [20]. Although live-cell imaging-based methods are extremely important when analyzing a phenomenon over time, the nature of these applications usually allows for the analysis of a limited number of cells, which may be misleading in regard to the total population. The second group comprises methods suitable for a larger number of cells. This group includes image analysis techniques, such as cis, medial, and trans Golgi imaging [21], direct stochastical optical reconstruction microscopy (dSTORM) [22], and flow cytometry methods, such as pulse-shape analysis (PulSA) [23, 24]. While these methods have more statistical power, they are either not readily applicable for the average user [21, 22] or lack spatial information [23, 24]. Therefore, we developed an imaging flow cytometry-based method for the analysis of Golgi fragmentation [25] that can overcome most of the disadvantages described above.

Imaging flow cytometry combines the information-rich morphological data acquisition of microscopy with the rapid, high-throughput quantification of flow cytometry [26–28]. Briefly, cells in suspension pass through a flow cell, where they are illuminated by a light-emitting diode (LED) and several lasers, resulting in a bright-field image, as well as up to 10 different fluorescent images obtained through different channels on the ImageStream (an Imaging Flow Cytometer by Amnis, part of Luminex Corporation). This is done at a rate of up to 5000 cells/s and, when coupled with the manufacturer's comprehensive image analysis software, IDEAS, it enables the rapid, high-throughput, unbiased quantification of both the quantity and morphological features of individual cells.

Our method for quantifying Golgi fragmentation is as simple as IF for any Golgi structural protein, and does not require any additional system establishment (e.g., stable overexpression). Hence, it can be easily transferred between systems. Moreover, given the large number of cells collected automatically by the ImageStream system, the results are robust, high-throughput, and statistically significant. In this chapter, we describe in a step-wise manner our method for the easy analysis of Golgi fragmentation by imaging flow cytometry.

2   Materials

2.1   Cell Culture and Stimulation

1. HeLa cells (ATCC).
2. 100 mm culture dishes.
3. 60 mm culture dishes.
4. Complete Dulbecco's Modified Eagle's medium (DMEM), supplemented with 10% fetal bovine serum (FBS), 100 U/mL penicillin, and 100 μg/mL streptomycin.
5. Trypsin C solution.
6. Sterile 1× phosphate-buffered saline (PBS).
7. Incubator with 5% CO_2 at 37 °C.
8. Nocodazole diluted in dimethyl sulfoxide (DMSO) to a 5 mM stock solution.
9. Brefeldin A (BFA) diluted in ethanol to a 0.7 mM stock solution.

2.2   Cell Collection and Fixation

1. Ice-cold, pure methanol (*see* **Note 1**).
2. Vortex mixer.
3. 15 mL conical centrifuge tube.
4. Benchtop centrifuge.
5. Incubator with 5% CO_2 at 37 °C.
6. Warm, sterile 1× PBS.
7. Ice-cold, sterile 1× PBS.

2.3   Immuno-fluorescence Labeling

1. Ice-cold 1× PBS.
2. Blocking / permeabilization buffer: 2% bovine serum albumin (BSA) and 0.1% Triton X-100 in 1× PBS.
3. Primary antibody solution: Primary rabbit antibody, either GM130 or Giantin (Abcam), diluted 1:100 in 1× PBS.
4. Intelli-Mixer RM-2l, supplied with 15 mL tubes adapter, set for 99° rotation angle.
5. Secondary antibody solution: AF568 anti-rabbit antibody (Invitrogen, diluted 1:500) together with DAPI (1 mg/mL, diluted 1:100) in 1× PBS.
6. 1.5 mL Eppendorf centrifuge tube.

2.4 Imaging Flow Cytometry and Image Analysis

1. ImageStream X Mark II imaging flow cytometer and software (INSPIRE).

2. IDEAS analysis software, preferably version 6.2 or higher.

3 Methods

3.1 Cell Culture and Stimulation

1. Grow HeLa cells (*see* **Note 2**) in a 100 mm culture dish in complete DMEM. Place dishes in a CO_2 incubator. Typically, cells should be passaged 2–3 times a week.

2. Plate 5×10^5 cells in a 60 mm culture dish; use one dish per condition. Keep two dishes for the non-treated and flow controls and another dish for the positive control (nocodazole or BFA, *see* **Note 3**). Grow the cells until they reach 70–80% confluency.

3. Add either nocodazole or BFA to the culture medium. If nocodazole is used, its final concentration should be 0.3 μM and then incubate cells for 4 h and 16 h (two different conditions). If BFA is used, its final concentration should be 0.2 μM and then incubate cells for 1 h (*see* **Note 4**).

3.2 Cell Collection and Fixation

1. Collect the media from each culture dish into a 15 mL conical centrifuge tube. Place the tube on ice (*see* **Note 5**).

2. Wash the culture dishes with a warm 1× PBS solution. Aspirate the PBS and add 0.5 mL Trypsin C solution to each dish. Incubate the dishes in an incubator for 3–5 min to allow cells to detach from the plate.

3. Use the media collected into a tube in **step 1** to wash the detached cells from the dish. Collect the media and washed cells and return to the tube.

4. Centrifuge cells for 5 min at $1000 \times g$.

5. Aspirate the supernatant add 5 mL of ice-cold 1× PBS and centrifuge again for 5 min at $1000 \times g$.

6. Aspirate the supernatant and then resuspend the cell pellet in 0.5 mL ice-cold 1× PBS.

7. Set the vortex to a low mixing level (2–3) and gently shake the cell suspension while dripping 1.7 mL ice-cold methanol into the tube.

8. Incubate samples overnight at 4 °C (*see* **Note 6**).

3.3 Immuno-fluorescence Labeling

1. Add 6 mL ice-cold 1× PBS to each tube and centrifuge for 5 min at $1000 \times g$. Repeat this step to remove any residual methanol.

2. Blocking / permeabilization: Aspirate the supernatant and resuspend the cell pellet in 0.5 mL blocking /permeabilization buffer. Incubate for 5 min at room temperature.

3. Add 6 mL ice-cold 1× PBS and centrifuge for 5 min at $1000 \times g$.

4. Primary antibody labeling: Aspirate the supernatant in the tube and resuspend the cell pellet in 100 μL primary antibody solution (*see* **Note 7**). Mix in Intelli-Mixer for 1 h at 4 °C.

5. Add 6 mL ice-cold 1× PBS and centrifuge for 5 min at $1000 \times g$.

6. Secondary antibody labeling: Aspirate the supernatant in the tube and resuspend the cell pellet in 100 μL secondary antibody solution. Mix in Intelli-Mixer for 30–60 min at 4 °C (*see* **Note 8**).

7. Add 6 mL ice-cold 1× PBS and centrifuge for 5 min at $1000 \times g$.

8. Aspirate the supernatant in the tube, resuspend the cell pellet in 200 μL 1× PBS, and then transfer to a clean 1.5 mL Eppendorf centrifuge tube.

3.4 Image Acquisition

1. Turn on the imaging flow cytometer, server, and acquisition computer. Log in to the acquisition computer (*see* **Note 9**).

2. Start INSPIRE. Once loaded, click on 'STARTUP' and make sure that all the calibrations and tests have passed (marked green).

3. Check the excitation/emission data for your dyes and turn on the relevant lasers and set them to maximal power. In the example we provide (Fig. 1), we used the 561 nm laser (200 mW) for the AF568 dye and 405 nm laser (120 mW) for the DAPI.

4. Turn on the relevant channels for your dyes. These should include the bright-field channels (#1 and #9 for a two-camera instrument) and side-scatter channel (#6 or #12, depending on your dye combination). In the example we provide, AF568 was acquired on channel 4 and DAPI was read on channel 7.

5. Load your first sample – start with the sample that contains all dyes, which is expected to produce the highest staining intensity.

6. Create a scatter plot of the Area vs. Aspect ratio of the bright-field channel. Gate on single cells according to your population distribution (*see* **Note 10**).

7. Create a histogram of Gradient_RMS of the bright-field channel. Draw a linear gate that includes values between 55 and 90, to collect only gated cells that are focused.

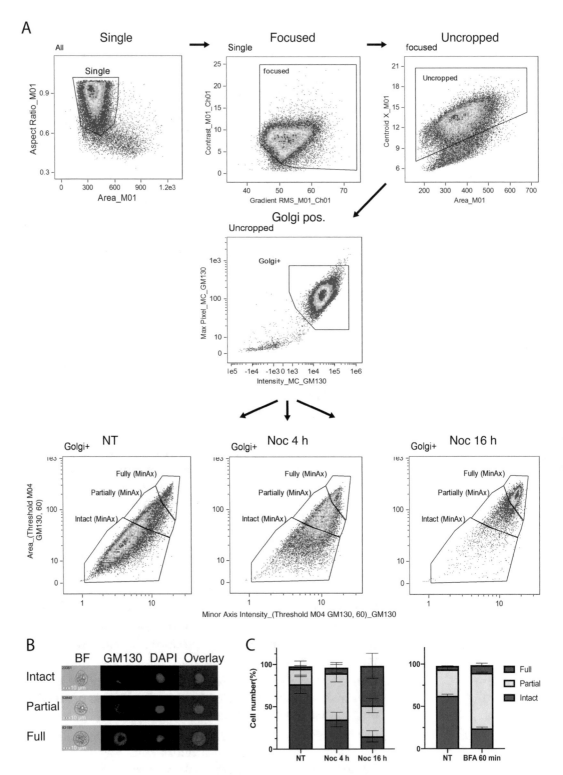

Fig. 1 Gating strategy for the analysis of Golgi fragmentation and expected results. (**a**) Flow chart scheme of the gating strategy for Golgi fragmentation analysis. (**b, c**) HeLa cells were treated with 0.3 μM nocodazole (Noc) for 4 h or 16 h, or treated with 0.2 μM BFA for 1 h; non-treated (NT) cells served as control. Cells were washed, fixed and stained for GM130 and DAPI as described in the protocol. (**b**) Representative image of an intact, partially fragmented and fully fragmented Golgi (BF- Bright-field, Red- GM130, Blue- DAPI). (**c**) Expected results for the treatment with nocodazole (left) or BFA (right). Results are shown as the percentage of cells from each population, and are an average of three independent experiments ± SEM

8. Create a 'Raw Max Pixel' histogram for each fluorescent channel, taken from the 'single' gate. Make sure the values do not reach 4095, which indicates the camera is saturated. If they do, reduce the relevant laser intensity until the values are within the target range.

9. Collect ~30,000 cells from each sample. If a rare population is assessed, collect enough cells so that the final population contains at least 500 cells.

10. Maintain focus and cell position during the acquisition. If the autofocus fails to maintain its focus, it could be disabled by going to Instrument → Advanced settings → Autofocus and disabling the 'autofocus' option. Once the autofocus is disabled, the focus can be manually changed on the 'Focus' tab.

11. If the cell position in the flow is not centralized, it can be adjusted manually by moving its position to 'Centering'.

12. After recording your samples, acquire single stained controls with the compensation wizard, using the same laser settings as those used for the experimental samples (**Note 11**).

3.5 Image Analysis

1. Start the IDEAS analysis software (preferably version 6.2 or higher).

2. Open the first file and calculate the compensation using the single-stain files acquired, with the compensation wizard in IDEAS.

3. Gate for single cells using the Area vs. Aspect ratio (Fig. 1a). If a cell-cycle analysis is required, make sure to include telophase cells, which are a close-to-doublet population (i.e., have higher area and lower aspect ratio).

4. Gate on focused cells using the Gradient RMS vs. Contrast of the bright-field image (Fig. 1a).

5. Gate on uncropped cells using the Area vs. Centroid X (Fig. 1a).

6. Gate on cells with a positive Golgi staining, by comparing the staining to control samples labeled with all the dyes except for the Golgi staining.

7. Using 'Tag Objects', manually select a group of cells with an intact Golgi and another group featuring a fragmented Golgi. Include around 100 cells per group.

8. Calculate several masks that delineate the Golgi staining. Usually, this would be done using the 'Morphology' and 'Threshold' masks, and we recommend using several 'Threshold' values. In the example provided, we used 'Threshold_60' (top 60% intensity pixels).

9. Calculate all the features using the 'Add Multiple Features' option in the 'Features' panel. Select all the feature options under the 'Size', 'Location', 'Shape', and 'Texture' categories. As signal strength may differ between samples and between experiments, it is undesirable to enable this feature option. Choose the masks you just created, choose the Golgi staining channel under 'Image', and click on 'Add Features'.

10. Add a new statistics table. Under 'Statistics', choose 'RD-Mean', and 'RD-Median'. Choose as a reference population the intact Golgi cells chosen in **step 8**, then click 'Add' to calculate.

11. Copy the values obtained to an Excel file and arrange them in descending order according to the RD values.

12. Examine the ability of the features that are at the top of the list to separate between the chosen populations.

13. Apply the selected features to the whole population and verify that they correspond to the different Golgi morphologies.

14. Save the template. Open the files recorded from the samples treated with nocodazole for 4 h and 16 h using this template.

15. It is recommended to merge the nocodazole-treated and control samples into one file, as it makes it easier to determine the gating. To do so, in each file click on Tools → Create Data File from Populations, and then select the single, focused, uncropped, Golgi-positive population, and create a *.CIF file. Then, click on 'Merge CIF files', select the files you just created, and use the template that was saved in the previous step.

16. Determine the gating that distinguishes between the different populations according to the controls (Fig. 1a).

17. Apply the gating on all the other experimental datasets.

18. If you have IDEAS version 6.3, quantification of the different Golgi morphologies could also be done by using the Machine Learning Wizard. In this case, use the populations chosen in **step 8** to allow the wizard to create a classifier that combines several features weighted for the best separation between populations. An example is provided in Fig. 2.

19. Once the Golgi fragmentation is determined, it could be combined with additional parameters or markers:

    (a) Cell cycle and division stages: Label DNA using a dye compatible with the ones you used to stain your samples. Common dyes and their corresponding channels (in brackets) include DAPI (Ch7), SyBR Green (Ch2), propidium iodide (Ch4), and Draq5 (Ch11). You can quantify the different stages according to the DNA fluorescence intensity, plotted as a histogram on a linear scale.

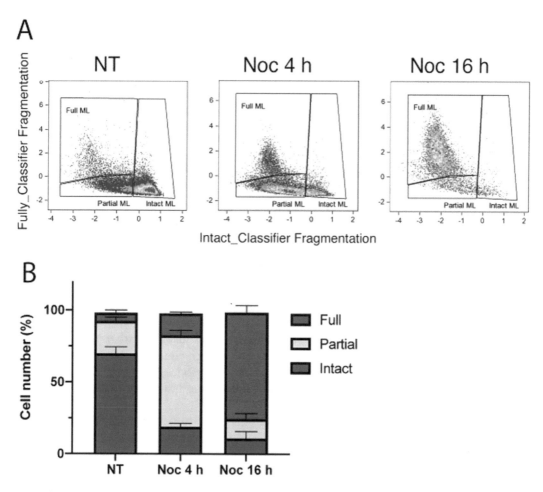

Fig. 2 Machine learning analysis of Golgi fragmentation. (**a**) Golgi fragmentation plot using the values of the classifier generated by the machine learning tool. (**b**) Cells were treated as described in Fig. 1. The expected results are shown as the percentage of cells from each population, and are an average of three independent experiments ± SEM

(b) Cell division: using the DNA staining, you can quantify the stages of cell division by DNA morphology (examples can be found in [25, 29]).

(c) Apoptosis: Apoptotic cells will have fragmented and condensed DNA. This can be quantified, for example, by using the contrast of the bright-field image vs. the area of the top 30% intensity pixels of the DNA staining [25].

(d) Co-localization: The co-localization of the Golgi with other proteins or organelles can be quantified by fluorescently labeling them and calculating the similarity using the similarity feature (i.e., log-transformed Pearson's Correlation Coefficient, a measure of the degree to which two images are linearly correlated within a masked region) or bright detail similarity (i.e., the intensity of localized bright spots whose radius is three pixels or less within the masked area in the image, with the local background around the spots removed) [29].

4   Expected Results

Typical results of the Golgi fragmentation in HeLa cells are shown in Fig. 1. Nocodazole treatment for 4 h increases the percentage of the partially fragmented Golgi population, and the 16 h treatment, the fully fragmented Golgi population is observed. BFA treatment results in rapid partial fragmentation of the Golgi (Fig. 1b, c).

An example of using the machine learning algorithm is shown in Fig. 2. It seems that this algorithm identifies more cells with totally fragmented Golgi following the nocodazole treatment compared to our previous analysis. This may be a result of the more accurate estimation made by machine learning, as it takes into account a combination of several features rather than only two. The advantages of our analysis pipeline are that it can be easily adapted to other experimental systems [30] and that it provides an unbiased, robust, high-throughput quantification of Golgi fragmentation.

5   Notes

1. Other fixatives may be used (e.g., PFA, ethanol), though they might induce some fragmentation. A preliminary experiment should be done to establish which fixation protocol works best for the specific cell line and antibody being used.

2. Cancer cell lines are often characterized by a fragmented Golgi structure. Therefore, the base-line populational distribution may be different for each cell line. To analyze the effect of a specific agent on Golgi fragmentation, it is recommended to choose a cell line in which the majority of the cells in population demonstrate an intact structure of the Golgi complex (e.g., HeLa).

3. As the Golgi structure and its dynamics might be cell-dependent (*see* **Note 2**), it is recommended to perform a preliminary experiment using nocodazole (4 h and 16 h) and BFA (1 h) to determine the border for the discrimination of each population (as shown in Fig. 1).

4. The treatment time should be determined empirically according to the cell type used. In many cell types, 4 h incubation with nocodazole will induce partial fragmentation, while 16 h will result in mitotic arrest and full Golgi fragmentation.

5. Treatment with nocodazole for 16 h induces mitotic arrest, which in many cell types results in the detachment of cells from the plate. Therefore, it is important to collect the suspension cells in the media.

6. Samples can be stored after fixation for up to 2 weeks at 4 °C.

7. GM130, Giantin, and GRASP65 antibodies were tested in our system [25]. Other Golgi markers may be used, and may be combined with other stainings.

8. Keep the second antibody solution in the dark until it is required for use and return to the dark after adding it to the cell pellet.

9. For a detailed explanation on how to use the instrument and analysis software, please refer to the user manual.

10. If you need to include in analysis the dividing cells in the telophase stage, increase the gate to include higher-area, low-aspect-ratio cells as well.

11. Single-cell controls intensity should be as bright as experimental samples.

References

1. Klute MJ, Melancon P, Dacks JB (2011) Evolution and diversity of the Golgi. Cold Spring Harb Perspect Biol 3:a007849. https://doi.org/10.1101/cshperspect.a007849

2. Gosavi P, Gleeson PA (2017) The function of the Golgi Ribbon structure – an enduring mystery unfolds! BioEssays 39:1700063. https://doi.org/10.1002/bies.201700063

3. Gardner BM, Pincus D, Gotthardt K, Gallagher CM, Walter P (2013) Endoplasmic reticulum stress sensing in the unfolded protein response. Cold Spring Harb Perspect Biol 5:a013169. https://doi.org/10.1101/cshperspect.a013169

4. Callegari S, Dennerlein S (2018) Sensing the stress: a role for the UPR(mt) and UPR(am) in the quality control of mitochondria. Front Cell Dev Biol 6:31. https://doi.org/10.3389/fcell.2018.00031

5. Davidson PM, Bigerelle M, Reiter G, Anselme K (2015) Different surface sensing of the cell body and nucleus in healthy primary cells and in a cancerous cell line on nanogrooves. Biointerphases 10:031004. https://doi.org/10.1116/1.4927556

6. Harbauer AB, Zahedi RP, Sickmann A, Pfanner N, Meisinger C (2014) The protein import machinery of mitochondria-a regulatory hub in metabolism, stress, and disease.

Cell Metab 19:357–372. https://doi.org/10.1016/j.cmet.2014.01.010

7. Ravichandran Y, Goud B, Manneville JB (2020) The Golgi apparatus and cell polarity: roles of the cytoskeleton, the Golgi matrix, and Golgi membranes. Curr Opin Cell Biol 62:104–113. https://doi.org/10.1016/j.ceb.2019.10.003

8. Ayala I, Mascanzoni F, Colanzi A (2020) The Golgi ribbon: mechanisms of maintenance and disassembly during the cell cycle. Biochem Soc Trans 48:245–256. https://doi.org/10.1042/BST20190646

9. Hicks SW, Machamer CE (2005) Golgi structure in stress sensing and apoptosis. Biochim Biophys Acta 1744:406–414. https://doi.org/10.1016/j.bbamcr.2005.03.002

10. Farber-Katz SE, Dippold HC, Buschman MD, Peterman MC, Xing M, Noakes CJ, Tat J, Ng MM, Rahajeng J, Cowan DM, Fuchs GJ, Zhou H, Field SJ (2014) DNA damage triggers Golgi dispersal via DNA-PK and GOLPH3. Cell 156:413–427. https://doi.org/10.1016/j.cell.2013.12.023

11. McKinnon CM, Mellor H (2017) The tumor suppressor RhoBTB1 controls Golgi integrity and breast cancer cell invasion through METTL7B. BMC Cancer 17:145. https://doi.org/10.1186/s12885-017-3138-3

12. Liu C, Mei M, Li Q, Roboti P, Pang Q, Ying Z, Gao F, Lowe M, Bao S (2017) Loss of the golgin GM130 causes Golgi disruption, Purkinje neuron loss, and ataxia in mice. Proc Natl Acad Sci U S A 114:346–351. https://doi.org/10.1073/pnas.1608576114

13. Thayer DA, Jan YN, Jan LY (2013) Increased neuronal activity fragments the Golgi complex. Proc Natl Acad Sci U S A 110:1482–1487. https://doi.org/10.1073/pnas.1220978110

14. Fujimoto T, Kuwahara T, Eguchi T, Sakurai M, Komori T (2018) Iwatsubo T (2018) Parkinson's disease-associated mutant LRRK2 phosphorylates Rab7L1 and modifies trans-Golgi morphology. Biochem Biophys Res Commun 495:1708–1715. https://doi.org/10.1016/j.bbrc.2017.12.024

15. Miyata S, Mizuno T, Koyama Y, Katayama T, Tohyama M (2013) The endoplasmic reticulum-resident chaperone heat shock protein 47 protects the Golgi apparatus from the effects of O-glycosylation inhibition. PLoS One 8:e69732. (2013). https://doi.org/10.1371/journal.pone.0069732

16. Casey CA, Thomes P, Manca S, Petrosyan A (2018) Giantin is required for post-alcohol recovery of Golgi in liver cells. Biomol Ther 8:150. https://doi.org/10.3390/biom8040150

17. Aistleitner K, Clark T, Dooley C, Hackstadt T (2020) Selective fragmentation of the trans-Golgi apparatus by Rickettsia rickettsii. PLoS Pathog 16:e1008582. https://doi.org/10.1371/journal.ppat.1008582

18. Hansen MD, Johnsen IB, Stiberg KA, Sherstova T, Wakita T, Richard GM, Kandasamy RK, Meurs EF, Anthonsen MW (2017) Hepatitis C virus triggers Golgi fragmentation and autophagy through the immunity-related GTPase M. Proc Natl Acad Sci U S A 114:E3462–E3471. https://doi.org/10.1073/pnas.1616683114

19. Colanzi A, Hidalgo Carcedo C, Persico A, Cericola C, Turacchio G, Bonazzi M, Luini A, Corda D (2007) The Golgi mitotic checkpoint is controlled by BARS-dependent fission of the Golgi ribbon into separate stacks in G2. EMBO J 26:2465–2476. https://doi.org/10.1038/sj.emboj.7601686

20. Tangemo C, Ronchi P, Colombelli J, Haselmann U, Simpson JC, Antony C, Stelzer EH, Pepperkok R, Reynaud EG (2011) A novel laser nanosurgery approach supports de novo Golgi biogenesis in mammalian cells. J Cell Sci 124:978–987. https://doi.org/10.1242/jcs.079640

21. Chia J, Goh G, Racine V, Ng S, Kumar P, Bard F (2012) RNAi screening reveals a large

22. Gunkel M, Erfle H, Starkuviene V (2016) High-content analysis of the Golgi Complex by correlative screening microscopy. Methods Mol Biol 1496:111–121. https://doi.org/10.1007/978-1-4939-6463-5_9

23. Ramdzan YM, Polling S, Chia CP, Ng IH, Ormsby AR, Croft NP, Purcell AW, Bogoyevitc MA, Ng D, Gleeson PA, Hatters DM (2012) Tracking protein aggregation and mislocalization in cells with flow cytometry. Nat Methods 9:467–470. https://doi.org/10.1038/nmeth.1930

24. Toh WH, Houghton FJ, Chia PZ, Ramdzan YM, Hatters DM, Gleeson PA (2015) Application of flow cytometry to analyze intracellular location and trafficking of cargo in cell populations. Methods Mol Biol 1270:227–238. https://doi.org/10.1007/978-1-4939-2309-0_17

25. Wortzel I, Koifman G, Rotter V, Seger R, Porat Z (2017) High throughput analysis of Golgi structure by imaging flow cytometry. Sci Rep 7:788. https://doi.org/10.1038/s41598-017-00909-y

26. Barteneva NS, Fasler-Kan E, Vorobjev IA (2012) Imaging flow cytometry: coping with heterogeneity in biological systems. J Histochem Cytochem 60:723–733. https://doi.org/10.1369/0022155412453052

27. Basiji D, Ortyn WE, Liang L, Venkatachalam V, Morrissey P (2007) Cellular image analysis and imaging by flow cytometry. Clin Lab Med 27:653–670. https://doi.org/10.1016/j.cll.2007.05.008

28. Zuba-Surma EK, Kucia M, Abdel-Latif A, Lillard JW Jr, Ratajczak MZ (2007) The Image-Stream System: a key step to a new era in imaging. Folia Histochem Cytobiol 45:279–290

29. Wortzel I, Hanoch T, Porat Z, Hausser A, Seger R (2015) Mitotic Golgi translocation of ERK1c is mediated by a PI4KIIIbeta-14-3-3gamma shuttling complex. J Cell Sci 128:4083–4095. https://doi.org/10.1242/jcs.170910

30. Eisenberg-Lerner A, Benyair R, Hizkiahou N, Nudel N, Maor R, Kramer MP, Shmueli MD, Zigdon I, Cherniavsky LM, Ulma A, Sagiv JY, Dayan M, Dassa B, Rosenwald M, Shachar I, Li J, Wang Y, Dezorella N, Khan S, Porat Z, Shimoni E, Avinoam O, Merbl Y (2020) Golgi organization is regulated by proteasomal degradation. Nat Commun 11:409. https://doi.org/10.1038/s41467-019-14038-9

Chapter 11

Flow Imaging of the Inflammasome: Evaluating ASC Speck Characteristics and Caspase-1 Activity

Abhinit Nagar and Jonathan A. Harton

Abstract

Examining inflammasome-associated speck structures is one of the most preferred and easiest ways to evaluate inflammasome activation. Microscopy-based evaluation of specks is preferable, but this approach is time-consuming and limited to small sample sizes. Speck-containing cells can also be quantitated by a flow cytometric method, time of flight inflammasome evaluation (TOFIE). However, TOFIE cannot perform single-cell analysis such as simultaneously visualizing ASC specks and caspase-1 activity, their location, and physical characteristics. Here we describe the application of an imaging flow cytometry-based approach that overcomes these limitations. Inflammasome and Caspase-1 Activity Characterization and Evaluation (ICCE) is a high-throughput, single-cell, rapid image analysis utilizing the Amnis ImageStream X instrument with over 99.5% accuracy. ICCE quantitatively and qualitatively characterizes the frequency, area, and cellular distribution of ASC specks and caspase-1 activity in mouse and human cells.

Key words Inflammasome, Caspase-1, Imaging Flow Cytometry

1 Introduction

Inflammasomes are multi-protein complexes comprised of a sensor that recruits and activates caspase-1 either directly or via an adaptor molecule, ASC [1, 2]. Inflammasome assembly is accompanied by the formation of a singular, toroidal, approximately 1 μm in diameter, perinuclear structure called a "speck". Since speck formation is a rapid event, it is a preferred read-out for inflammasome activation. Apart from speck formation, caspase-1 activation is another direct read-out for inflammasome activation [3, 4]. Caspase-1 activation is measured by immunoblotting for cleaved caspase-1 or measuring its cleavage product IL-1β or by detection of binding by the fluorescent caspase-1 inhibitor FLICA [5, 6]. In contrast, speck formation is quantitated by microscopy- and/or flow cytometer-based methods. Only microscopy-based techniques allow single-cell analysis of speck formation. However, microscopy is time-consuming,

Natasha S. Barteneva and Ivan A. Vorobjev (eds.), *Spectral and Imaging Cytometry: Methods and Protocols*,
Methods in Molecular Biology, vol. 2635, https://doi.org/10.1007/978-1-0716-3020-4_11,
© Springer Science+Business Media, LLC, part of Springer Nature 2023

limited to small sample sizes, and subject to user bias for determination of relevant structures versus artifacts [7, 8]. Moreover, subjective biases in microscopy are amplified further when quantitative evaluation of speck size and distribution or patterns of caspase-1 activity are included.

Inflammasome and caspase-1 activity characterization and evaluation (ICCE) is an automated, high-throughput, highly precise computational and quantitative single-cell imaging flow cytometry-based approach that enumerates ASC speck-containing cells, evaluates speck size, and the localization of caspase-1 activity. High-throughput quantitative imaging allows users to simultaneously perform single-cell and population-based analyses, a degree of evaluation that was previously impossible. While several other AMNIS-based methods for quantifying specks are available, none employ a masking strategy to discriminate ASC-containing aggregates in general from structures that meet the established criteria for size and shape of the speck. ICCE uses a masking strategy that distinguishes between and quantifies cells with individual specks and those without while simultaneously assessing caspase-1 activity. Thus, ICCE also allows users to study changes in speck characteristics along with localization of caspase-1 activity. Finally, the masking/gating strategy described in this chapter can be further customized to accommodate additional measures including speck circularity, distance between multiple speck-like structures, and areas of caspase-1 activity.

2   Materials

2.1   Cells

1. Human kidney epithelial cells (HEK239T).
2. human monocytic cell line (THP-1).
3. Immortalized bone marrow-derived macrophages (iBMDMs) (kind gift from Dr. Kate Fitzgerald, UMass Medical School, Worcester, MA).
4. Primary human monocytes (University of Nebraska Medical Center).

2.2   Cell Culture

1. Dulbecco's Modified Eagle Medium/high glucose (DMEM).
2. Fetal bovine serum (FBS).
3. $1\times$ Glutamax.
4. Human AB Serum.
5. RPMI-1640.
6. β-mercaptoethanol.
7. Phosphate Buffered Saline (PBS), $1\times$, 0.0067 M PO_4, without Calcium, Magnesium.

2.3 Reagents	1. Nigericin.
	2. ATP.
	3. FLICA-660 caspase-1 assay kit.
	4. Ethylenediaminetetraacetic acid solution (EDTA).
	5. Phosphate-buffered saline (PBS) 10×, 0.067 M PO4, without calcium and magnesium.
	6. AMNIS cell suspension buffer prepared by dissolving 0.5 mM EDTA in 1× PBS. Prepare the solutions using ultrapure water.
	7. 4% paraformaldehyde (PFA).
	8. FuGENE6.
	9. Lipopolysaccharide (LPS) (O26:B6).
	10. TritonX-100.
	11. Fish gelatin.
	12. Bovine serum albumin (BSA).
	13. Rabbit anti-ASC (N15)-R.
	14. Mouse anti-ASC (D-3).
	15. Alexa Fluor®488 goat-anti-rabbit IgG.
	16. Alexa Fluor®594 goat-anti-rabbit IgG2a.
	17. 1× Trypsin-EDTA.
	18. DAPI.

2.4 Instrument and Software

1. Amnis ImageStream X (MilliporeSigma, Inc.) equipped with 405, 488, and 642 nm lasers with a single camera (six channels).
2. Acquisition software – INSPIRE.
3. Data analysis software – Image Data Exploration & Analysis Software (IDEAS).

2.5 Expression Plasmids and DNA Transfection

1. Expression plasmids encoding human caspase-1 [9] and human NLRP3 [10] have been previously described.
2. The plasmid encoding GFP-ASC was generated by cloning ASC into the pEGFP-C3 expression vector backbone after HindIII and KpnI digestion.

3 Methods (Fig. 1)

3.1 Inflammasome Reconstitution and Activation

HEK293T cells:

1. Plate HEK293T cells (2×10^5/well) in 12-well plates (Corning) in 1 mL of DMEM supplemented with 10% fetal bovine serum (FBS) and 1× Glutamax and incubate at 37 °C with 5%

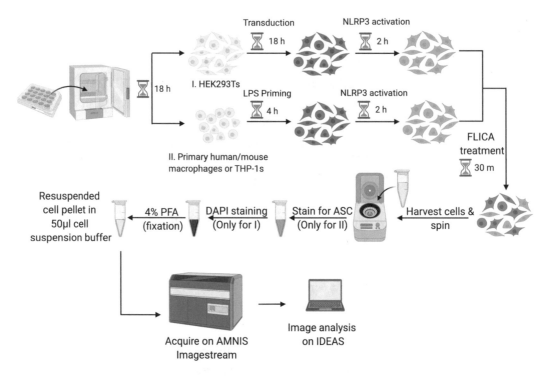

Fig. 1 Schematic representation of workflow

CO_2. Cells can also be cultured in antibiotic (Penicillin/Strep-
tomycin)-containing media.

2. After 18–20 h, transfect cells with plasmids encoding pro-cas-
pase-1 (20 or 50 ng) and GFP-ASC (50 ng) with or without
NLRP3 (100 ng) and incubate at 37 °C with 5% CO_2 (transfection volume 50 µL) (*see* **Notes 1** and **2**).

3. At approximately 18 h after transfection, add 50 µL of 110 µM
nigericin (final concentration 5 µM) or vehicle for 2 h.

4. After 2 h, set the pipettor to 950 µL and gently aspirate media
from the wall of the well while tilting the plate. The volume left
in each well will be approximately 150 µL.

5. Freshly prepare FLICA 660 stock solution by adding 4 µL
FLICA to 30 µL serum-free DMEM media. Add 30 µL of
this working FLICA solution to each well (1:45 final dilution)
and incubate for 30 min at 37 °C with 5% CO_2 (*see* **Note 2**).

6. Aspirate media gently from each well while titling the plate and
gently wash cells twice by adding 200 µL of 1× wash buffer
(supplied with the FLICA kit) along the side of the well. Gently
aspirate the wash buffer completely after the last wash.

7. Add 50 µL 1× trypsin-EDTA per well and let sit at room
temperature for 30–60 s. Resuspend cells with 1 mL culture
media and transfer to 1.5 mL Eppendorf tubes. Centrifuge at

500g for 5 min, discard the supernatant and wash the cell pellet with 1 mL of 1× PBS and spin down at 500g for 5 min. Completely remove supernatant and fix cells by adding 200 µL of 4% paraformaldehyde and incubate for 15 min at room temperature.

8. Centrifuge cells at 500g for 5 min and aspirate the supernatant. Proceed with DAPI staining (See Preparation of samples for acquisition by the Amnis ImageStream X).

3.2 THP1 Cells

1. Place THP1 cells at 1 × 10^6/well in 6-well plates in 1 mL of RPMI supplemented with 10% FBS, 1×-β-mercaptoethanol, and 1×-Glutamax, and incubate at 37 °C with 5% CO_2. Cells can also be cultured in antibiotic (Penicillin/Streptomycin)-containing media.

2. After 18–20 h, prime cells with LPS by adding 200 µL of a 1 µg/mL stock solution of LPS (O26:B6) to each well (final concentration of 100 ng/mL LPS) and incubate for 3–4 h.

3. Following LPS treatment, stimulate cells by adding 300 µL of serum-free RPMI-containing 50 µM nigericin or 25 mM ATP to each well. Incubate at 37 °C with 5% CO_2 for 30 min.

4. Centrifuge the cells at 500g for 5 min and wash cell pellets with 1× PBS. Resuspend cells in 200 µL serum-free RPMI-containing FLICA660 (1:45 final dilution) and incubate for 30 min at 37 °C with 5% CO_2.

5. Centrifuge the cells at 500g for 5 min, remove wash buffer, and resuspend cells in 1 mL FLICA-free media and incubate at 37 °C with 5% CO_2 for 1 h (see **Note 3**).

6. Centrifuge the cells at 500g for 5 min, aspirate supernatant and wash cell pellet with 500 µL of 1× PBS and spin down at 500g for 5 min.

7. Aspirate the PBS and fix cells by resuspending the cell pellet in 100 µL of 4% paraformaldehyde (PFA) and incubate for 15 min at room temperature. Centrifuge the cells at 500g for 5 min.

8. Aspirate fixative. Permeabilize by resuspending the cells in 100 µL of 0.1% TritonX-100 and incubate for 10 min at room temperature. Centrifuge the cells at 500g for 5 min.

9. Aspirate permeabilization buffer. Add 500 µL of PBS blocking buffer containing 5% fish gelatin, 1% BSA, and 0.05% Triton X-100 and incubate for 1 h at room temperature. Centrifuge the cells at 500g for 5 min.

10. Aspirate the blocking buffer. Resuspend the pellet in 250 µL of wash buffer (PBS containing 1% fish gelatin, 1% BSA, and 0.5% Triton X-100) containing Rabbit anti-ASC (N15)-R (1:250) or Mouse anti-ASC (D-3) (1:500) (Santa Cruz). Stain cells in the antibody solution at room temperature for 2 h.

11. Centrifuge the cells at 500g for 5 min and wash cell pellets 3× with 500 μL of wash buffer. Centrifuge the cells at 500g for 5 min and aspirate the supernatant.

12. Add 500 μL of wash buffer containing secondary antibody, Alexa Fluor®488 goat-anti-rabbit IgG (1:500) or Alexa Fluor®594 goat-anti-rabbit IgG$_{2a}$ (1:1000) and incubate covered with foil, at room temperature for 1 h.

13. Spin cells at 500g for 5 min and aspirate the supernatant. Proceed with acquisition (see **ImageStream X acquisition parameters**).

3.3 Immortalized BMDMs (iBMDMs): Wild-type and Caspase-1/11$^{-/-}$

1. Plate iBMDMs cells at 1×10^6/well in 6-well plates in 2 mL DMEM supplemented with 10% FBS, and 1×-Glutamax and incubate at 37 °C with 5% CO_2. Cells can also be cultured in antibiotic (Penicillin/Streptomycin)-containing media.

2. Follow **steps 2** and **3**, as described for THP1 cells.

3. After 30 min, aspirate the media leaving 150 μL and make up the volume to 200 μL with serum-free DMEM-containing FLICA660 (1:45 final dilution) and incubate for 30 min at 37 °C with 5% CO_2. Follow **steps 4** and **5** described for HEK293T cells.

4. Aspirate media from each well and pour 1 mL of serum-free DMEM carefully along the sides of the wall of the well. Incubate for 1 h instead of washing with 1× wash buffer.

5. Aspirate media and treat cells with 50 μL trypsin-EDTA per well and spin down at 500g for 5 min. Wash cell pellet with 1× PBS and spin down at 500g for 5 min.

6. Follow **steps 6–12**, as described above for THP1 cells.

7. Spin cells at 500g for 5 min and aspirate the supernatant. Proceed with DAPI staining (see Preparation of samples for acquisition by the Amnis ImageStream®^{X}).

3.4 Primary Human Monocytes (See Note 4)

1. Plate primary human cells at 1×10^6/well in 6-well plates in 1 mL hDMEM containing 100 ng/mL LPS (O26:B6) and incubate at 37 °C with 5% CO_2 for 3–4 h. Cells can also be cultured in antibiotic (Penicillin/Streptomycin)-containing media.

2. Following LPS treatment, stimulate cells with 5 mM ATP and 10 μM nigericin in 1 mL of hDMEM for 30 min, making the volume 2 mL/well.

3. Follow **steps 4–13** as described above for THP-1 cells.

3.5 Preparation of Samples for Acquisition by the Amnis ImageStream X

1. Resuspend cells in 100 μL of DAPI solution (1 μg/mL in 1× PBS supplemented with 0.5 mM EDTA (ISXII cell suspension buffer)) for 10 min at room temperature (RT) to stain nuclei (see **Note 5**).

2. Wash cells once with 500 μL of 1× PBS and resuspend cells in 50 μL ISXII cell suspension buffer by gently tapping the tube (see **Note 6**).

3. The samples are now ready to be acquired using the ISXII.

3.6 ImageStream X Acquisition Parameters: (See Note 7)

1. Acquire samples on the ImageStream X using INSPIRE® software.

2. Set flow rate to minimum and objective magnification to 60× (0.33 μm per pixel resolution) for all samples.

3. Load a multi-fluorophore labeled sample (positive control) on the AMNIS and adjust the laser settings with *raw max pixel* using *area* to ensure visualization of a cell. This will avoid fluorophore over-saturation. Laser settings used for acquisition are shown in Table 1 (see **Note 8**).

4. To ensure collection of single-cell events, gate for cells in the default bright field mask using *Area* versus *Aspect Ratio* (see **Note 9**).

5. Draw a region (R1) (0–50 for area and 0–1 for aspect ratio) to gate for debris. Draw two other regions, R2 (area 70–600; aspect ratio 0.5–1) and R3 (area 100–300; aspect ratio 0.6–1), to gate for all cells and the singlet population respectively. Select R2 population as the collection population and record 10,000 events in R3 for every sample.

6. To acquire samples for compensation controls, turn off Ch04 (Bright field) and 785 nm laser (Side-scatter; SSC) leaving the setting for other channels as used for acquisition of all the samples. Acquire 1000 events for all single-stained samples (see **Notes 2** and **3**). All samples and controls are acquired as raw image files (.rif) (see **Note 10**).

Table 1
AMNIS instrument laser power setting used

Emission channel	Fluorophore	Excitation laser (nm)	Laser power (mW)
1	DAPI	405	10
2	GFP	488	10
3	Blank	–	–
4	Bright field	–	–
5	FLICA	642	150
6	Side scatter (SSC)	785	2

Channel 1 is used for DAPI, Channel 2 is used for GFP, Channel 3 is turned off for acquisition, Channel 4 s used for bright field, Channel 5 is used for FLICA, and Channel 6 is used for side-scatter

Image analysis using Image Data Exploration & Analysis Software (IDEAS):

3.7 Compensation and Data Analysis Template Generation on IDEAS

1. Load the .rif files of single-stained samples as individual compensation controls into the compensation wizard in IDEAS® (*see* **Note 11**).

2. Follow instructions to finalize the compensation matrix. Once the compensation matrix is finalized, click "Finish" to save the compensation matrix file (.ctm). This .ctm file will be used for the rest of the analysis.

3. Apply the compensation matrix to a positive control (e.g., Nigericin-treated sample) raw data file (.rif) to generate both a compensated image file (.cif) and a corresponding data analysis file (.daf).

4. Create all the masks and required gates (as described below) on the .daf file of positive control sample (*see* **Note 12**).

5. After completing the masking and gating on the positive control file, generate a data analysis template (.ast) (*see* **Note 13**).

6. The template (.ast) along with the compensation matrix is applied to the rest of the experimental samples using the multiple file batch tool in IDEAS (*see* **Note 13**).

Masking Strategy

Mask names are *italicized*, and mask features are ***bold italicized***.

Focused Cell and Nucleus Mask

1. Apply an *object mask* to *default bright field mask* (M04) to accurately calculate the area of the cell and improve ***Gradient_RMS*** of the mask (*see* **Note 14**).

2. Similarly, apply a *morphology mask* to default channel 1 mask (M01) to accurately calculate the area of nucleus.

3. Recalculate ***area, gradient_RMS,*** and ***aspect ratio*** features on these new masks for both brightfield and nucleus channel.

ASC Mask

Details of the individual masks used in developing the ASC speck mask are shown in Table 2.

1. Create two different *spot masks* ASC speck 1 and ASC speck 2 from the *default system mask* for GFP (M02) (*see* **Note 15**). ASC speck 1 and ASC speck 2 are set with different levels of stringency for parameters of spot-to-cell count and area (*see* **Note 16**).

2. Use the Boolean operator OR to combine the Speck 1 and Speck 2 masks to create a new mask, ASC speck 3.

Table 2
Masking strategy for ASC speck (*see* Note 19)

Mask name	Reconstituted ASC specks	Native ASC Specks	
	HEK293T (ASC-GFP)	THP-1	Primary human monocytes
ASC Mask 1	Spot (M02, Ch02-GFP-ASC, Bright, 2, 10, 3)	Spot (M02, Ch02-AF488, Bright, 3, 5, 1)	Spot (M05, Ch05-AF594, Bright, 3, 5, 1)
ASC Mask 2	Spot (M02, Ch02-GFP-ASC, Bright, 1, 20, 3)	Spot (M02, Ch02-AF488, Bright, 3, 3, 1)	Spot (M05, Ch05-AF594, Bright, 3, 3, 1)
ASC Mask 3	ASC Mask 1 or ASC Mask 2	ASC Mask 1 or ASC Mask 2	ASC Mask 1 or ASC Mask 2
ASC Mask 4	LevelSet (ASC Mask 3, Ch02-GFP-ASC, Middle, 5)	LevelSet (ASC Mask 3, Ch02-AF488, Middle, 5)	LevelSet (ASC Mask 3, Ch05-AF594, Middle, 5)
ASC Mask 5	LevelSet (ASC Mask 3, Ch02-GFP-ASC, Bright, 5)	LevelSet (ASC Mask 3, Ch02-AF488, Bright, 5)	LevelSet (ASC Mask 3, Ch05-AF594, Bright, 5)
ASC Mask 6	ASC Mask 4 or ASC Mask 5	ASC Mask 4 or ASC Mask 5	ASC Mask 4 or ASC Mask 5
ASC Mask 7	Fill (ASC Mask 6)	Fill (ASC Mask 6)	Fill (ASC Mask 6)
ASC Mask 8	Range (ASC Mask 7, 15–500, 0.4–1)	Range (ASC Mask 7, 0–200, 0.3–1)	Range (ASC Mask 7, 0–200, 0.5–1)
ASC Mask 9	Intensity (ASC Mask 8, Ch02-GFP-ASC, 750–4095)	Intensity (ASC Mask 8, Ch02-AF488, 300–4095)	Intensity (ASC Mask 8 Ch05-AF594, 750–4095)
ASC Mask 10	Dilate (ASC Mask 9, 1)	Dilate (ASC Mask 9, 1)	Dilate (ASC Mask 9, 1)
ASC Mask 11	Fill (ASC Mask 10)	Fill (ASC Mask 10)	Fill (ASC Mask 10)
ASC Mask 12	Threshold (ASC Mask 11, Ch02-GFP-ASC, 80)	Threshold (ASC Mask 11, Ch02-AF488, 52)	Threshold (ASC Mask 11, Ch05-AF594, 52)
ASC Mask 13	Range (ASC Mask 12, 15–450, 0.6–1)	Range (ASC Mask 12, 2–18, 0.5–1)	Range (ASC Mask 12, 5–20, 0.6–1)
ASC Mask 14	Component (1, Area, ASC Mask 13, Descending)	Component (1, Area, ASC Mask 13, Descending)	Component (1, Area, ASC Mask 13, Descending)

3. Create two different *levelset masks* (ASC speck 4 and ASC speck 5) and combine into a single mask (ASC speck 6) using the Boolean OR operation (*see* **Note 17**).

4. *Levelset mask* function in this masking strategy generates non-contiguous masks due to low pixel intensities. Thus, variations in the ASC speck 6 mask should be eliminated by using the fill function (ASC speck 7).

5. To differentiate between large atypical ASC aggregates and specks, apply the *range mask* to the ASC speck 7 mask to select for events with an area between 15 and 500 pixels and an aspect ratio of 0.4–1 resulting in the ASC speck 8 mask.

6. To minimize the background, apply the *intensity mask* function to ASC speck 8. Set the intensity threshold to 750–4095 to eliminate smaller non-speck-like aggregates (*see* **Note 18**). The maximum intensity is 4095. The new mask is named ASC Speck 9.

7. The size of mask should be adjusted to match the size of the speck using *dilate* and *fill masks* to generate ASC speck 10 and 11 masks respectively.

8. To adjust the stringency and better adjust the mask, apply *threshold mask* function to eliminate artificially small aggregate signals (ASC speck 12).

9. Apply range mask function to ASC speck 12 with the setting of area of 15–450 and aspect ratio of 0.6–1 to select for round specks (ASC speck 13).

10. Despite this elaborate masking strategy, some non-speck-like structures (smaller than expected size) will be selected. To eliminate any non-speck-like structure, rank ASC speck 13 into individual component masks using the *component mask* function. This mask function will rank ASC speck 13 individual masks by area, sorted from highest to lowest (component 1 = largest area). Any structure too large to be defined as the speck should be eliminated by ASC speck 8–13.

11. Select the *component mask* with largest area (component 1 mask) and designate it as ASC speck 14. ASC speck 14 is the mask that identifies specks of the expected size.

Active Caspase-1/FLICA Mask

Details of individual masks are shown in Table 3.

1. Apply *spot mask* function to default FLICA mask to differentiate between diffused and aggregate staining pattern (FLICA-I).

2. The FLICA-I mask should be further refined using the *intensity mask* function to set the lower intensity threshold (*see* **Note 20**). This mask is referred to as FLICA-II in Table 3.

3. To eliminate further small aggregates, use *erode mask* function to create FLICA-III mask.

Table 3
Masking strategy for FLICA aggregates

Mask name	Function	Setting
FLICA-I	Spot	Spot (M05, Ch05-FLICA, Bright, 5, 5, 2)
FLICA-II	Intensity	Intensity (FLICA-I, Ch05-FLICA, 150–4095)
FLICA Spot-III	Erode	Erode (FLICA-II, 1)
FLICA-IV	Dilate	Dilate (FLICA-III, 3)
FLICA-V	Range	Range (FLICA-IV, 20–1000, 0.6–1)

4. The *erode* function will reduce the size of the mask compared to actual FLICA aggregate. Apply the *dilate* function to adjust the mask size to match the active caspase-1 aggregates to allow area calculations. This mask is called FLICA-IV (*see* **Note 21**).

5. To select FLICA spots with nearly circular shape, apply *range mask* on FLICA-IV to select for events with an area between 20 and 1000 pixels and an aspect ratio of 0.6–1. This mask is called FLICA-V (*see* **Note 22**).

Gating Strategy

1. Plot **Gradient RMS** for brightfield (Ch4) and **gradient RMS** for DAPI and draw a region to gate both brightfield and nuclear images that were in focus. The gate can be named "Cells" (*see* **Note 23**).

2. To identify the single-cell population (Single BF), gate the "Cells" population for specific **area** (50–300) and **aspect ratio** (0.6–1) (*see* **Note 24**).

3. To select for the population with optimal nuclear staining (DAPI), gate "Single BF" based on **intensity** of nuclear stain.

4. To eliminate cells where cytokinesis is just starting, gate "DAPI" population for the **area** and **aspect ratio** of the nucleus mask to include cells containing single nuclei. This mask is designated "Single_DAPI" (*see* **Note 25**).

5. To segregate populations based on fluorophore colors, plot **mean pixel** for GFP and FLICA staining. These populations can be segregated into GFP+, FLICA+, Double-positives, Double-negatives.

Calculation of ASC Speck and FLICA Features

These calculations use the ASC speck 14 and FLICA-V masks: In IDEAS, click Analysis > Features. Select **spot count, area,** and **diameter** features for ASC speck 14 mask on Ch-02 GFP channel for calculations of the number, area, and diameter of ASC specks. Similarly, select **spot count** and **intensity** feature for FLICA-V mask on Ch-05 FLICA for calculations of the number and intensity of FLICA-stained spots.

1. Calculating the number and frequency of cells containing ASC specks: To identify cells with a speck, click "histogram plot" to plot the *spot count* of ASC speck mask 14 on the double-positive cells (GFP+ and FLICA+), which selects cells with a speck (Speck-positive cells; spot count = 1) and discriminates them from cells lacking a speck (Speck-negative cells; spot count = 0) (Fig. 2).

2. Calculating ASC speck area and diameter: To calculate speck area, apply *area* feature on ASC speck mask 14 on Ch-02 (GFP channel). Click "histogram plot" to plot the *area* of ASC speck mask 14 for speck-positive cell population (Fig. 3).

3. Calculating the number and frequency of FLICA-positive cells: The frequency of double-positive cells is a direct measure of FLICA-positive cells with substantial ASC staining (Fig. 4a).

4. Calculating the number and frequency of FLICA-aggregate-containing cells: To identify cells containing FLICA aggregates, apply *spot count* feature on FLICA-V on Ch-05 (FLICA) to the double-positive population (GFP+ and FLICA+). Click "histogram plot" to plot the *spot count* of FLICA-V mask on speck-positive cells. Cells with a diffuse distribution of FLICA staining will return a spot count = 0. Cells containing FLICA aggregates will have a spot count ≥ 1 (Fig. 4b, c).

5. Calculating the intensity of FLICA signal: To measure the extent of caspase-1 activation, plot (Histogram) the *intensity* of FLICA in double-positive cells. The median fluorescent intensity of FLICA can be calculated from the graph and can be included in the statistics report Fig. 4d).

Generating a Template File (.ast) for Batch Analysis

1. Once the masking and gating strategy are finalized, a statistics report can be generated (Reports>Generate Statistics Report). The choice of statistics may vary between users and should be determined as per the user's requirements. The steps to generate statistics report can be found in the IDEAS® user manual (*see* **Note 14**). Save the file as a template file (.ast).

2. To perform the batch analysis (Tools>Batch data files), select files for all the samples (.rif) as well as the compensation file (.cst) and the template file (.ast), then click submit to start the batch analysis. A visual description of the steps to perform a batch analysis can be found in the IDEAS®user manual (*see* **Note 14**).

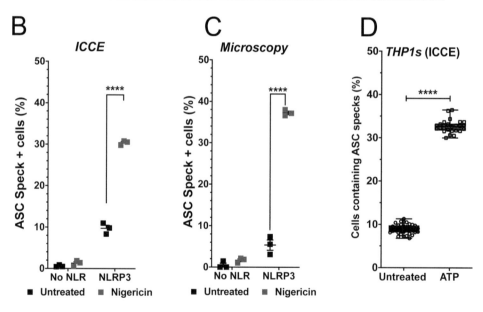

A

		ICCE				
		Exp-1	Exp-2	Exp-3	Mean	SD
No NLR	Untreated	0.48	0.44	0.86	0.59	0.23
	Nigericin	1.39	1.85	0.58	1.27	0.64
NLRP3	Untreated	8.36	9.82	10.99	9.72	1.32
	Nigericin	29.77	30.54	30.80	30.37	0.53

		Microscopy				
		Exp-1	Exp-2	Exp-3	Mean	SD
No NLR	Untreated	1.43	0.00	0.00	0.48	0.83
	Nigericin	1.02	1.85	2.11	1.66	0.57
NLRP3	Untreated	5.55	7.40	3.07	5.34	2.17
	Nigericin	38.00	36.58	37.14	37.24	0.72

Fig. 2 Quantification of specks positive cells in HEK293Ts and THP1s cells. (**a–d**) Comparison of ICCE and microscopy-to quantitate speck-positive cells. (**a, b**) Tabular representation of three independent experiments. (**c**) Frequency of speck-positive cells in nigericin treated inflammasome reconstituted HEK293T cells measured by ICCE. (**d**) Frequency of speck-positive cells in nigericin treated inflammasome reconstituted HEK293T cells measured by microscopy. (**e**) Frequency of speck-positive cells in nigericin treated LPS-primed THP1s cells measured by ICCE

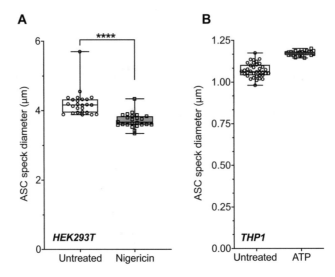

Fig. 3 Measurement of ASC speck diameter in HEK293Ts and THP1s cells using ICCE. (**a**) Speck diameter measured by ICCE in inflammasome reconstituted HEK293Ts cells either treated with 5 μM nigericin or vehicle. (**b**) Speck diameter measured by ICCE in LPS primed THP-1 cells either treated with 5 mM ATP or vehicle for 30 min

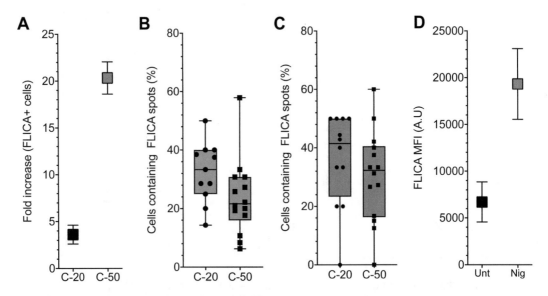

Fig. 4 Evaluation of caspase-1 activation in HEK293T cells. HEK293T cells were transfected with 100 ng NLRP3, 50 ng GFP-ASC and 20 or 50 ng of capsase-1 (C-20 or C-50), treated with 5 μM nigericin, and stained for active caspase-1 using FLICA-660, as described. (**a**) Fold-increase in FLICA-positive cells. (**b**) Cells containing organized FLICA aggregates (Measure using spot count for FLICA-V mask). (**c**) ASC-speck-positive cells containing organized FLICA aggregates. (**d**) FLICA intensity in cells treated with nigericin (in C-20 cells only)

4   Notes

1. Maintain DNA concentration at 1 μg/well for each transfection using pcDNA3 (empty vector). We find that 2.5 μL of FuGENE6 per μg of DNA is optimal although other transfection reagents may be useful.

2. To prepare a single stain control for the FLICA channel, transfect cells as described but use an untagged-ASC expression plasmid instead of GFP-ASC. Stain with FLICA as described for other samples. For a single stain control for GFP, transfect cells as described but do not add FLICA.

3. Incubating THP-1 cells in serum-free media is an alternative to washing with FLICA wash buffer and is less stressful to cells. To prepare single stain controls, stain one sample with FLICA only and one sample with ASC only.

4. Primary human monocytes were obtained from University of Nebraska Medical Center and cultured in DMEM supplemented with 10% Human AB Serum (hDMEM).

5. For THP1 cells, ASC localizes to nuclei [11] and thus DAPI staining should be avoided to prevent interference with ASC staining. The gating strategy for THP1 and hMDMs are described in detail in our manuscript [12]. Briefly, plot *gradient_RMS* for the brightfield channel and gate for cells in focus (Cells) and identify "Singlets" by plotting *Aspect ratio* versus *Area* for the bright field channel. The cell population positive for ASC staining can be selected by plotting *intensity* or *mean pixel value* for channel specific for secondary fluorophore of ASC staining (If Alexa fluor 488 is used then Ch-02, and if Alexa fluor 594 is used then Ch-05).

6. Use of a pipette or vortex mixer should be avoided at this stage, because vortexing or pipetting can cause adherence of cell to Eppendorf tube/pipette tip walls resulting in cell loss.

7. An Amnis ImageStream X (MilliporeSigma, Inc.) equipped with 405, 488, and 642 nm lasers with a single camera (six channels) was used to acquire experimental samples using the INSPIRE® software.

8. Determine the saturation of each fluorophore in its corresponding channel by plotting the Raw Max pixel feature versus the Area feature of the default mask of that channel. A saturated signal is determined by a value of 4000 of raw max pixel value. An optimal range of max pixel value should be 500–3000.

9. Aspect ratio feature measures circularity, thus helping to distinguish between singlets, doublets, and cell clumps when used together with the Area feature.

10. The instrument acquires all sample and control images as raw image files (.rif).

11. All samples and compensations are to be analyzed using the IDEAS software.

12. Masks are defined regions of interest that are computationally calculated by INSPIRE. A mask defines a specific region of an image which can be used for specific feature calculations.

13. Refer to the IDEAS manual to understand different file types and how to save a template (.ast) file. The IDEAS manual can be downloaded from "https://www.luminexcorp.com/down load/amnis-ideas-software-user-manual/"

14. Gradient RMS measures the average contrast of the image. A high gradient RMS represents better-focused images.

15. A spot mask differentiates between punctate and diffuse staining.

16. ASC speck 1 accommodates smaller aggregates, whereas ASC speck 2 allows for detection of multiple aggregates (using spot-to-cell count).

17. A levelset mask identifies pixels in non-homogeneous regions into three different settings: dim, middle, and bright. Use ASC speck 3 mask as the input for a levelset mask. The levelset function with the middle and bright setting should be used to identify fairly bright and bright spots respectively.

18. The lower limit of the intensity range is to be determined by staining pattern. The lower limit selected should eliminate small non-speck like ASC-GFP aggregates.

19. For native ASC specks, which are smaller than GFP-ASC specks, do not alter the order of the masks. However, settings can be fined tuned to provide a better fit for speck characteristics (Table 2).

20. The lower limit of the intensity range is to be determined by staining pattern. The lower limit selected should eliminate less organized tiny FLICA aggregates. In our hands, this value was approximately 150, but will vary depending on staining intensity.

21. The parameters for *dilate* feature should be adjusted visually to match the size of FLICA aggregates. We found that a parameter of 3 pixels was optimal.

22. Double-positive samples with diffuse FLICA staining will have a spot count of zero on the FLICA-V mask.

23. The gate limits should be determined by manually visualizing cells from each side of the gate. A schematic of the gating strategy can be obtained from Nagar et al. [12].

24. Area settings will define the size of the cells acquired and the aspect ratio defines circularity and can exclude irregularly shaped cells or doublets. These parameters if set appropriately gate for single cells.

25. Cells where cytokinesis is just starting will have a dividing nucleus but will be included as single cells. Evaluating the area and aspect ratio of the DAPI-stained nuclei will allow elimination of cells with dividing nuclei (karyokinesis approaching completion and cytokinesis just starting) that would otherwise be counted as singlets because of their low aspect ratio in the nucleus mask.

Acknowledgements

We thank Drs. Bibhuti Mishra and Kate Fitzgerald for their kind gift of immortalized BMDMs. We would also like to thank Richard De Marco for his training and instruction in using AMNIS IDEAS and assisting in the evaluation of our masking strategy. We would also like to thank the American Association of Immunologists (AAI) for the opportunity to present this method at IMMUNOLOGY 2017, Washington, D.C., May 12–16, 2017 [13]. Figure 1 was generated using Biorender (www.Biorender.com).

References

1. Broz P, Dixit VM (2016) Inflammasomes: mechanism of assembly, regulation and signalling. Nat Rev Immunol 16:407–420

2. Latz E, Xiao TS, Stutz A (2013) Activation and regulation of the inflammasomes. Nat Rev Immunol 13:397–411

3. Sester DP, Thygesen SJ, Sagalenko V et al (2015) A novel flow cytometric method to assess inflammasome formation. J Immunol 194:455–462. https://doi.org/10.4049/jimmunol.1401110

4. Stutz A, Horvath GL, Monks BG, Latz E (2013) In: De Nardo C, Latz E (eds) ASC speck formation as a readout for inflammasome activation, Methods in Molecular Biology, vol 1040. Humana Press, Totowa, pp 91–101. https://doi.org/10.1007/978-1-62703-523-1_8

5. Martinon F, Burns K, Tschopp J (2002) The inflammasome: a molecular platform triggering activation of inflammatory caspases and processing of proIL-beta. Mol Cell 10:417–426

6. Sokolovska A, Becker CE, Ip WK et al (2013) Activation of caspase-1 by the NLRP3 inflammasome regulates the NADPH oxidase NOX2 to control phagosome function. Nat Immunol 14:543–553

7. Arrasate M, Finkbeiner S (2005) Automated microscope system for determining factors that predict neuronal fate. Proc Natl Acad Sci U S A 102:3840–3845

8. Carter M, Shieh JC (2015) Guide to research techniques in neuroscience. Elsevier Academic Press, Amsterdam

9. Atianand MK et al (2011) Francisella tularensis reveals a disparity between human and mouse NLRP3 inflammasome activation. J Biol Chem 286:39033–39042

10. O'Connor W, Harton JA, Zhu S et al (2003) Cutting edge: CIAS1/cryopyrin/PYPAF1/NALP3/CATERPILLER 1.1 is an inducible inflammatory mediator with NF-kappa B suppressive properties. J Immunol 171:6329–6333

11. Bryan NB, Dorfleutner A, Rojanasakul Y, Stehlik C (2009) Activation of inflammasomes requires intracellular redistribution of the apoptotic speck-like protein containing a caspase recruitment domain. J Immunol 182:3173–3182

12. Nagar A, DeMarco RA, Harton JA (2019) Inflammasome and caspase-1 activity characterization and evaluation: an imaging flow cytometer-based detection and assessment of inflammasome specks and caspase-1 activation. J Immunol 202:1003–1015

13. Nagar A, DeMarco RA, Harton JA (2017) A novel method of assessing ASC specks and caspase-1 activity by imaging flow cytometry. J Immunol 198(1 Suppl):64.2–64.2

Chapter 12

Quantitative Analysis of Latex Beads Phagocytosis by Human Macrophages Using Imaging Flow Cytometry with Extended Depth of Field (EDF)

Ekaterina Pavlova, Daria Shaposhnikova, Svetlana Petrichuk, Tatiana Radygina, and Maria Erokhina

Abstract

The existing methods of quantitative analysis of phagocytosis are characterized by a number of limitations. The usual method of manually counting phagocytosed objects on photographs obtained by confocal microscopy is very labor-intensive and time-consuming. As well, the resolution of conventional flow cytometry does not allow the fluorescence detection of a large number of phagocytosis objects. Thus, there is a need to combine the rapid analysis by flow cytometry and the visualization capability by confocal microscopy. This is possible due to imaging flow cytometry. However, until now, no protocols have allowed one to quantify phagocytosis at its high intensity. The present paper presents the developed and tested algorithm for assessing the level of phagocytic activity using flow cytometry with visualization and IDEAS software.

Key words Imaging flow cytometry, Phagocytosis, Extended Depth of Field

1 Introduction

Phagocytosis is one of the essential properties of macrophages and a fundamental process in innate and adaptive immunity [1]. Macrophages, as professional phagocytes, are able to engulf a large number of biological or artificial objects. In most cases, conventional flow cytometry or light microscopy, especially confocal laser scanning microscopy (CLSM), is used to quantify the activity of the phagocytic process [2–4].

Quantification of phagocytosis activity based on light microscopy takes a long time to obtain enough images for further statistical analysis and requires skills in using complex software for automatic image processing. Flow cytometry does not have such disadvantages, but a significant limitation of flow cytometry is cell autofluorescence, which does not allow accurate identification of

Natasha S. Barteneva and Ivan A. Vorobjev (eds.), *Spectral and Imaging Cytometry: Methods and Protocols*,
Methods in Molecular Biology, vol. 2635, https://doi.org/10.1007/978-1-0716-3020-4_12,
© Springer Science+Business Media, LLC, part of Springer Nature 2023

the phagocytic population. Imaging flow cytometry (IFC) has greater fluorescence sensitivity than conventional flow cytometry of smaller objects like bacteria [5]. In this regard, the use of the advantages of IFC for the study of phagocytosis activity is a relevant methodological approach.

Many authors have used IFC to analyze the activity of phagocytosis in various subpopulations of immune cells [6–11]. The advantages of IFC are evident for the quantification of a small number of phagocytosed objects when the cell autofluorescence histogram significantly overlaps with the fluorescence of a small number (1–3) of phagocytosed objects [11]. But the accurate counting of a large number of phagocytosed fluorescent objects (>5) in cells is still a matter of methodological discussion.

Thus, Ploppa et al. [6] concluded that during phagocytosis of *S. aureus* by neutrophils using the «Spot mask» only up to 5 bacteria per cell can be distinguished separately. Pelletier et al. [7] demonstrated that using a «Spot mask» on average, only 2.6 individual phagocytized targets per cell could be resolved at 20× magnification and no more than 13 phagocytosed targets per cell at 60× magnification.

Since «Spot mask» does not allow determination of a large number of objects, the strategy for assessing high phagocytic activity is usually based on the analysis of bacterial fluorescence intensity. Unfortunately, most authors using this strategy don't indicate how the fluorescence intensity parameter correlates with the number of phagocytosed objects. Park and co-authors (2020) described a strategy for the quantitative detection of fluorescent microplastic bead particles inside phagocytic cells based on the assumption that fluorescence intensity is proportional to the number of beads taken up by each phagocytic cell. In another study, using conventional flow cytometry to analyze the number of phagocytosed latex particles, the authors suggested that each fluorescence intensity peak on the histogram corresponds to one engulfed latex particle [12].

In the current protocol, we propose a more accurate method for quantifying the activity of phagocytosis based on the preliminary analysis of the intensity of fluorescent particles without cells. Understanding the distribution of the particle fluorescence intensity makes it possible to identify gates for a subpopulation of macrophages with low (on average, up to 10 particles), medium (from 10 to 20), and high (more than 20) phagocytosis activity. We validated our method by comparing the obtained results with the confocal laser scanning microscopy (CLSM) method.

As the CLSM method showed when implementing receptor phagocytosis through FcRs the proportion of cells that engulf 1–10 IgG particles (low) was 85.9% after 3 h and 48.2% after 6 h of incubation. Analysis of the "high" subpopulation, which engulfs more than 20 particles, showed that its proportion increased from 2.3% to 33.6% after 3 and 6 h of incubation, respectively. A feature of phagocytosis through FcRs is the formation of clusters containing more than 20 particles after 6 h of incubation. Analysis of the data using the IFC method revealed similar dynamics. The percentage of cells in the "low" group decreased from 78.9 to 58.4, and the percentage of cells in the "high" group increased from 4.3 to 18.8 between 3 and 6 h of incubation with IgG particles, respectively. Thus, both methods confirm significant activation of receptor phagocytosis after 6 h of phagocytosis of opsonized particles.

We determined that when analyzing high phagocytic activity by IFC, it is not possible to create a suitable mask to distinguish between individual particles due to their low contrast in the cell. The Enhanced Depth Field (EDF) option can be used to increase the image contrast of small fluorescent objects in cells. Typically, EDF mode is used in FISH assays to count probes on chromosomes. EDF mode has about ten times less spread in the values of focus-sensitive features compared to standard visualization [13]. EDF makes it easy to create a mask for counting spots and get a plot of the distribution of the number of particles in a cell. But it is not recommended to use the integrated intensity histogram due to the distortion of the intensity of individual particles introduced when using the EDF option. EDF collapses multiple focal plane images into a single plane, making it difficult to distinguish between internalized objects and cell-bound objects [14]. Some approaches, such use of additional antibodies to phagocytized objects [9] or the use of dye conjugates fluoresce only in phagosomes [10, 11], make it possible to distinguish with high accuracy cell-bound objects from internalized ones, thereby making it possible to use the EDF option. This method analyzed macrophages characterized by very high activity of phagocytosis and engulfed a large number of particles, so we do not distinguish between objects associated with the cell surface from intracellular particles. But for cells with low phagocytosis activity, the above approaches may be necessary.

The approach we propose for the quantitative analysis of phagocytosis activity is quite simple to use and does not require much time; it does not require the use of complex masks; allows identify groups of macrophages with different activities of phagocytosis. This approach can be used to identify cell populations with high phagocytic activity, in which it is quite difficult to count individual phagocytosed objects.

2   Materials

2.1   Reagents

1. Complete culture medium: RPMI 1640 with 2 mM L-glutamine and 40 μg/mL gentamicin sulfate and 10% fetal bovine serum.

2. EDTA-Na$_2$ (Ethylenediaminetetraacetic acid disodium salt dehydrate) (Sigma-Aldrich, USA).

3. 37 wt. % in H_2O formaldehyde solution (Sigma-Aldrich, USA).

4. Fluoresbrite® BB Carboxylate Microspheres 1.00 μM were used as objects of phagocytosis (Polyscience, USA). Further in the text named "BB Carboxylate Microspheres" or latex particles.

5. IgG from human serum (Sigma-Aldrich, USA).

6. EDAC (1-Ethyl-3-(3-dimethylaminopropyl) carbodiimide, Sigma-Aldrich, USA).

7. PMA (Phorbol 12-myristate 13-acetate), ×1000 stock solution in DMSO (dimethyl sulfoxide).

8. PBS (Phosphate-buffered solution) pH 7.4.

9. Aqueous Mounting Medium with DAPI (Abcam, USA).

10. Sodium azide (powder, Sigma-Aldrich, USA).

2.2 Laboratory Equipment and Accessories

1. CO_2 incubator for culturing cells at 37 °C in a humidified 5% CO_2/95% air atmosphere/.

2. 6-well tissue-treated plates for macrophages and non-treated T-25 flasks for suspension culture.

3. 1.5 mL microcentrifuge tubes, serological pipets, and pipettes with tips.

4. Cell scrapers for gentle detachment of macrophage may be required.

5. Tabletop centrifuge.

6. Automated cell counter.

7. Amnis® ImageStreamX Mk II (Amnis-Luminex, USA) equipped with 375 nm, 405 nm, 488 nm, and 785 nm (side scatter) lasers, 60× magnification objective, and extended depth of field (EDF) option. The EDF function is required to accurately distinguish particles for Spot count.

8. IDEAS® version 6.2. analysis software (Amnis-Luminex, USA).

2.3 Cell Culture

Monocytic cell line THP-1 was purchased from the Type Culture Collection (Institute of Cytology, Russian Academy of Sciences (RAS), St-Petersburg, Russia).

3 Methods

3.1 Cells and Macrophage Differentiation Protocol

1. Cells were cultivated in a complete medium with densities below 1×10^6 per ml in T-25 flasks at 37 °C in a CO_2 incubator with 5% CO_2.

2. For macrophage differentiation, cells were counted and seeded in 6-well plates at a density of 0.2×10^6 cells per ml in complete

media, a total 5 mL per well. Add PMA to a final concentration 100 nM. Place the plate in a CO_2 incubator for 72 h. (*see* **Note 1**). Put a small piece of coverslip in each well if you plan to compare the data obtained by IFC with CLSM.

3.2 Phagocytosis Assay Technique

1. Conjugation of the latex particles with IgG from human serum was performed according to the manufacturer's protocol using EDAC for covalent conjugation (Technical data sheet 238C, Polyscience, USA). The stock solution of conjugated particles (~1% concentration) in PBS with 1 mM sodium azide is stored at 4 °C for no more than 3 months. The conjugated particles were added directly to the culture medium with ratio of cells to particles approximately 1:100.

2. In order to avoid clumping of the microspheres, sonicate the stock solution in a sonicating bath for 10 min and then vortex for 30 s. Add particles (×1000) to the complete fresh medium without FBS (*see* **Note 2**) to a final concentration of 0.001% and mix.

3. Aspirate supernatant and wash macrophages twice with sterile PBS to remove non-adherent cells.

4. Add 5 mL of fresh media without FBS with phagocytic particles and incubate the cells for 3 or 6 h in a CO_2 incubator.

5. Place plates with cells on ice to stop the phagocytic process.

6. Pull out the pieces of the coverslip, drop fixing buffer (*see* **Note 3**) to cover the slide completely, and after 10 min of incubation, wash the cells twice in PBS. Follow mounting medium instructions for mounting coverslips. We recommend using aqueous mounting media with DAPI.

7. Aspirate 4 ml of medium and add detachment stock solution (×100) directly to the well. Incubate for 10–15 min. If the cells don't detach, use a cell scraper (*see* **Note 4**).

Detachment Solution Stock solution 250 mM EDTA (×100) solution was prepared by adding 18.6 g EDTA to 150 mL deionized H_2O, adjusting to pH 6.14 with HCl as described [15]. The stock was filtered with a 0.2 µm nylon sterile syringe filter and stored at room temperature. The working EDTA solution was prepared by diluting the stock 250 mM EDTA solution to 15 mM in deionized H_2O. A working solution was added directly to the culture medium.

8. Harvest cells into 1.5 mL microcentrifuge tube, spin down 10 min at 800 × g at 4 °C, and discard the supernatant.

9. Add 1 ml fixative buffer and vortex cells, incubate 10 min at 4 °C, and centrifuge 10 min at 1000 × g. Discard the supernatant and resuspend the cell pellet in 1 Ml PBS. Cells can be stored at 4 °C in the dark for up to 24 h before analysis.

Fixative Buffer A stock 37 wt. % in H$_2$O formaldehyde solution (Sigma-Aldrich), diluted in PBS (pH 7.2–7.4) to a final concentration of 4%.

3.3 Data Acquisition and Analysis

1. Imaging cytometry analysis is performed on Amnis® ImageStreamX Mk II (Amnis-Luminex, USA) equipped with a 60× magnification objective and extended depth of field (EDF) option. Collect at least 5000 single cell events with the following laser settings: 405 nm = 175 mW (*See* **Note 5**) and 785 nm = 1 mW with the EDF option and without. Also, using the same instrument settings, record images of individual particles without cells. Collect bright-field images in Channel 01, Side Scatter in Channel 06, and images of phagocytic particles in Channel 07 (Fig. 1). Use the slow speed of flow rate for high-resolution imaging.

2. Analysis of raw data file (*.cif) was performed with IDEAS 6.2 (Amnis, EMD Millipore). In this experiment, it is not required to create a compensation matrix (*.ctm) as only one fluorochrome (BB Carboxylate Microspheres) is used.

3. To start data analysis, graph Bright field size vs. Aspect ratio using a linear scale to eliminate debris and cell aggregates. Create gate «Single cells». The Aspect ratio is the ratio of the image's long axis to its short axis. Accordingly, the closer the value of this parameter to 1, the more rounded shape the cell has. Based on visual control, we selected a cell population with Aspect Ratio_M01 > 0.75 and Area_M01 from 150 to 750 AU (arbitrary units) for further analysis.

4. In the next step, graph «Gradient RMS_M07_Ch07» for gating the phagocytic particles that are in focus. This step is especially necessary if the phagocytosis activity is analyzed

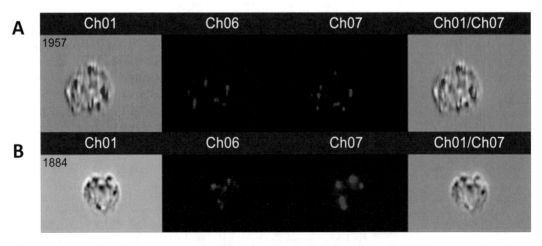

Fig. 1 Representative fluorescent cell image with EDF option (**a**) and without (**b**) EDF. Image gallery for a macrophage with phagocytic particles. Ch01: Bright-field; Ch06: SSC; Ch07: Fluorescent particles – BB Carboxylate Microspheres 1.00 μM; Ch01/Ch07: composite image

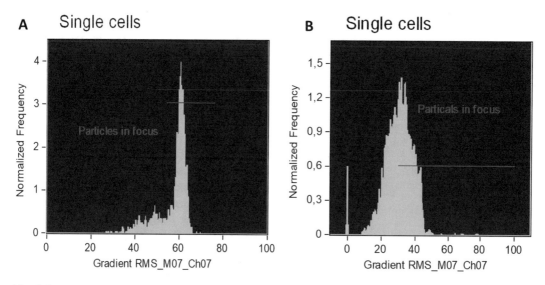

Fig. 2 Representative histograms of Gradient RMS_Ch07 (fluorescence of the phagocytic particles) with EDF (**a**) and without (**b**)

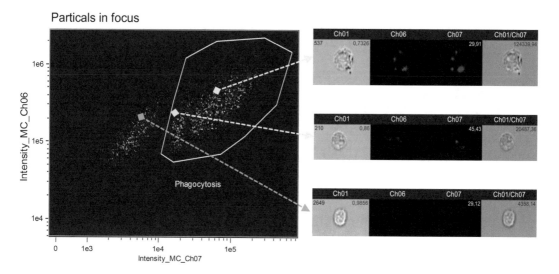

Fig. 3 The scatterplot for identification of a phagocytic population of macrophages and presented images of dots in different scatterplot cellular populations

without the EDF option. EDF increases the particle contrast required to distinguish spots further accurately (Fig. 2).

For data without EDF we used Gradient RMS_M07_Ch07 > 30 AU

5. In order to discriminate the phagocytic population, use scatterplot Intensity_Ch07 vs. Intensity_Ch06 (Side scatter channel) (see Fig. 3). Two populations of cells are clearly distinguishable on the graph. Gating accuracy was verified by visual inspection (*see* **Note 6**).

6. Using a file with a collection of only particles without cells, determine the fluorescence intensity of a single particle. The gating strategy for particles is the same as for cells. Use scatterplot Aria vs. Aspect Ratio Bright Field to identify individual particles. We used the following parameters: Aria 0–170 and Aspect Ratio – 0 – 1. Next, graph Gradient RMS Ch07 and select the particles in focus.

Analysis of the latex particle fluorescence intensity showed that the fluorescence of a single latex particle corresponds to a wide range of fluorescence intensities (see Fig. 4a).

For a small number of objects in the image, it is possible to use the particle count mask (Spot) without the EDF option. We created the following "Spot mask": (Peak (M07, Ch07, Bright, 2.5), Ch07, 125-4095). Further analysis showed that the 1st peak on the histogram corresponds not to one particle

A   Beads in focus_M1

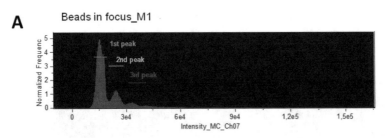

Intensity_MC_Ch07

Population	%Gated	Mean	Median	Std. Dev.	CV	Minimum	Maximum
Beads in focus_M1 & Single...	100	20130,54	16868,9	9056,75	44,99	4627,77	150909,73
1st peak & Beads in focus_...	64,5	15741,61	15646,03	1593,13	10,12	12615,13	19434,43
2nd peak & Beads in focus_...	18,6	24203,68	24129,63	1904,12	7,867	20512,61	28409,49
3rd peak & Beads in focus_...	6,34	34872,32	34271,47	2750,52	7,887	31000,29	40999,96

B   1st peak:

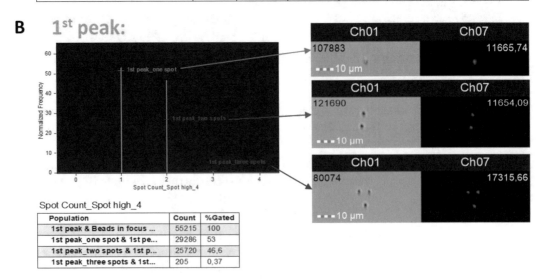

Spot Count_Spot high_4

Population	Count	%Gated
1st peak & Beads in focus ...	55215	100
1st peak_one spot & 1st pe...	29286	53
1st peak_two spots & 1st p...	25720	46,6
1st peak_three spots & 1st...	205	0,37

Fig. 4 (a) The histogram of particle intensity distribution. (b) Examples of particles in the 1st peak of the histogram of spot count particles distribution

but 1–2, and, in a small percentage of cases, three particles. (see Fig. 4b). Thus "one peak ≠ one particle". But each peak on the intensity histogram increases the range by +1–2 particles.

7. The median fluorescence intensity of the first peak is 15,646 arbitrary units (AU) (CV = 10.12), the second peak is 24,129 AU (CV = 7.87), and the third – 34,271 AU (CV = 7.87). Thus, each peak is approximately +10,000 AU in the median, and + 1/2 particles. So, using these data, we created the following gates to identify subpopulations of macrophages with different phagocytosis activity: Low phago – from 13,300 to 124,600* AU (up to ~10 particles), Medium phago – from 124,600 to 247,000 (up to ~20 particles) and High phago – more than 247,000 (over 20 particles) (see Fig. 5). Get statistics on the percentage of macrophages in each gate.

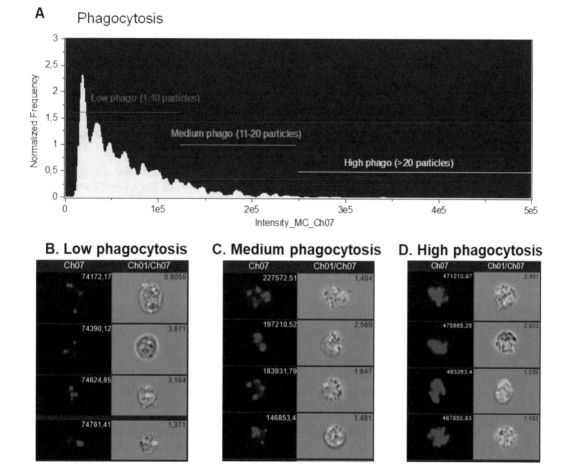

Fig. 5 The histogram of the phagocytized particles intensity distribution is used for gating of macrophages with different phagocytosis activity (a). Representative images of macrophages with phagocytosed particles from gates with Low (b), Medium (c), and High (d) activity of phagocytosis

Single cells_focused particles

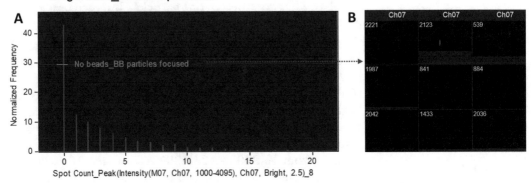

Fig. 6 Histogram of Spot count distribution of BB-carboxylate particles; **(a)** and gallery of images in gate «No particles_BB particles focused» **(b)**

8. Open *.cif file with EDF mode and create the following Mask to identify individual particles in the cell: Peak(Intensity(M07, Ch07, 1000-4095), Ch07, Bright, 2.5). Then use this Mask to create Feature Spot count: Spot Count_Peak(Intensity(M07, Ch07, 1000-4095), Ch07, Bright, 2.5)_8 (*See* **Note 7**). Using the EDF option simplifies the creation of a mask to detect individual particles in the cells.

9. Make sure that the no-particle gate («No particles_BB particles focused») does not contain or contains only rare images of cells with particles. (Fig. 6).

10. Check the accuracy of Spot count via the Image gallery (Fig. 7). Count the percentage in each spot (*see* **Note 8**).

4   Notes

1. Treatment of monocytic cells with THP-1 with 100 nM PMA for 72 h induces a pro-inflammatory macrophage phenotype characterized by very high activity of FcγR-mediated phagocytosis and a high level of secretion of pro-inflammatory cytokines [16].

2. In various protocols, it is recommended starving cells without serum for at least 1 h before the experiment. The presence of serum in the medium can reduce phagocytosis activity and can also contribute to particle agglutination [17].

3. If you wish to use additional staining with phalloidin for actin, remember that commercially available paraformaldehyde solutions contain methanol for stabilization. Methanol fixation is not compatible with phalloidin staining.

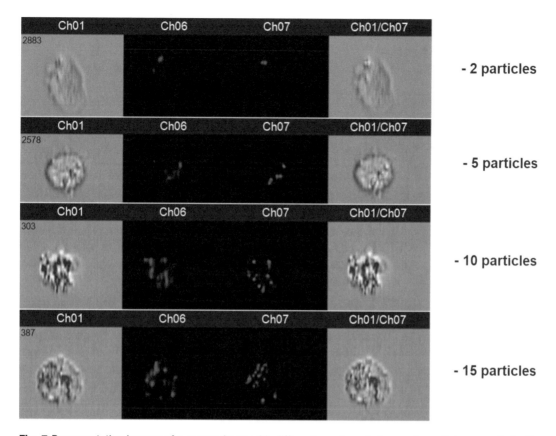

Fig. 7 Representative images of a macrophage with different number of phagocytic particles recorded with EDF option. Ch01: Bright-field; Ch06: SSC; Ch07: Fluorescent particles – BB Carboxylate Microspheres 1.00 µM; Ch01/Ch07: composite image

4. EDTA not only reduces cell damage during detachment but also prevents further formation of agglomerates in a cell suspension. In addition, rinsing cell suspension in high concentrations of EDTA reduces the number of particles non-specifically binding to the cell surface.

5. According to the manufacturer, latex particles have the following spectral characteristics: 360 nm/407 nm (Excitation/Emission). Despite this, in this experiment, a 405 nm laser was used to excite latex particles rather than 365 nm (maximum power 70 mV). The choice of the 405 nm laser for particle excitation was due to the fact that the 405 nm laser has higher power, allowing you to get a more intense signal from the particles. Median fluorescence intensity by 375 nm laser – 3668 AU vs. 13,656 AU by 405 nm laser. (Fig. 8).

6. Various strategies are used to gate a population of phagocytic cells. The most common is a graph of Intensity vs. Max Pixel or Raw Max Pixel [11, 18]. The parameter of side scattering reflects the intrinsic complexity of the cells; accordingly, this

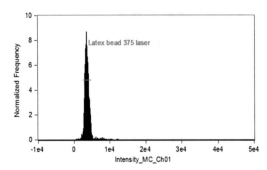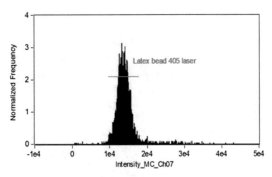

Fig. 8 Histograms of particle fluorescence intensity upon excitation by 375 nM and 405 nM lasers

value should increase as the number of phagocytosed objects in the cell increases. We have found that the graph of intensity of particle fluorescence vs intensity side scatter channel is more convenient.

7. Depending on the specific characteristics of the particles and the parameters of the experiment, it may be necessary to change the parameters of this mask. In order to do this, plot a histogram of particle fluorescence intensity in the "Single cells" gate and, using the gallery of images pre-sorted by Ch07 fluorescence intensity, check that the gate without particles ("No particles_BB focused particles") contains no or only single images of cells with particles.

8. Don't use an intensity histogram to assess phagocytosis activity due to the distortion of the intensity of individual particles introduced when using the EDF option.

Acknowledgments

The work was supported by the Russian Foundation for Basic Research, No 20–34–90161 project. We are also grateful for Daria Potashnikova (Lomonosov Moscow State University) for valuable advice and Ivan Vorobjev (School of Sciences and Humanities, Nazarbayev University) for help with the editing of the manuscript.

References

1. Rosales C, Uribe-Querol E (2017) Phagocytosis: a fundamental process in immunity. Biomed Res Int 2017:9042851. https://doi.org/10.1155/2017/9042851

2. Steinkamp JA, Wilson JS, Saunders GC, Stewart CC (1982) Phagocytosis: flow cytometric quantitation with fluorescent microspheres. Science (New York, NY) 215(4528):64–66. https://doi.org/10.1126/science.7053559

3. Caponegro MD, Thompson KK, Tayyab M, Tsirka SE (2020) A rigorous quantitative approach to Analyzing phagocytosis assays. Bio-protocol 10(15):e3698. https://doi.org/10.21769/BioProtoc.3698

4. Sattler N, Monroy R, Soldati T (2013) Quantitative analysis of phagocytosis and phagosome maturation. Methods Mol Biol 983:383–402.

https://doi.org/10.1007/978-1-62703-302-2_21

5. Basiji DA (2016) Principles of amnis imaging flow cytometry. In: Barteneva N, Vorobjev I (eds) Imaging flow cytometry, Methods in molecular biology, vol 1389. Humana Press, New York. https://doi.org/10.1007/978-1-4939-3302-0_2

6. Ploppa A, George TC, Unertl KE, Nohe B, Durieux ME (2011) ImageStream cytometry extends the analysis of phagocytosis and oxidative burst. Scand J Clin Lab Invest 71:362–369. https://doi.org/10.3109/00365513.2011.572182

7. Pelletier MG, Szymczak K, Barbeau AM, Prata GN, O'Fallon KS, Gaines P (2017) Characterization of neutrophils and macrophages from ex vivo-cultured murine bone marrow for morphologic maturation and functional responses by imaging flow cytometry. Methods 112:124–146. https://doi.org/10.1016/j.ymeth.2016.09.005

8. Smirnov A, Solga MD, Lannigan J, Criss AK (2020) Using imaging flow cytometry to quantify neutrophil phagocytosis. Methods Mol Biol 2087:127–140. https://doi.org/10.1007/978-1-0716-0154-9_10

9. Smirnov A, Solga MD, Lannigan J, Criss AK (2017) High-throughput particle uptake analysis by imaging flow cytometry. Curr Protoc Cytom 80:11.22.1–11.22.17. https://doi.org/10.1002/cpcy.19

10. Xu C, Lo A, Yammanuru A, Tallarico ASC, Brady K, Murakami A, Barteneva N, Zhu Q, Marasco WA (2010) Unique biological properties of catalytic domain directed human anti-CAIX antibodies discovered through phage-display technology. PLoS One 5:e9625

11. Park Y, Abihssira-García IS, Thalmann S, Wiegertjes GF, Barreda DR, Olsvik PA, Kiron V (2020) Imaging flow cytometry protocols for examining phagocytosis of microplastics and bioparticles by immune cells of aquatic animals. Front Immunol 11:203. https://doi.org/10.3389/fimmu.2020.00203

12. Daigneault M, Preston JA, Marriott HM, Whyte MK, Dockrell DH (2010) The identification of markers of macrophage differentiation in PMA-stimulated THP-1 cells and monocyte-derived macrophages. PLoS One 5:e8668. https://doi.org/10.1371/journal.pone.0008668

13. Ortyn WE, Perry DJ, Venkatachalam V, Liang L, Hall BE, Frost K, Basiji DA (2007) Extended depth of field imaging for high speed cell analysis. Cytometry A 71:215–231. https://doi.org/10.1002/cyto.a.20370

14. Jenner D, Ducker C, Clark G, Prior J, Rowland CA (2016) Using multispectral imaging flow cytometry to assess an in vitro intracellular Burkholderia thailandensis infection model. Cytometry A 89:328–337. https://doi.org/10.1002/cyto.a.22809

15. Kaur M, Esau L (2015) Two-step protocol for preparing adherent cells for high-throughput flow cytometry. BioTechniques 59:119–126. https://doi.org/10.2144/000114325

16. Kurynina AV, Erokhina MV, Makarevich OA, Sysoeva VY, Lepekha LN, Kuznetsov SA, Onishchenko GE (2018) Plasticity of human THP-1 cell phagocytic activity during macrophagic differentiation. Biochem Biokhimiia 83:200–214. https://doi.org/10.1134/S0006297918030021

17. Golovkina MS, Skachkov IV, Metelev MV, Kuzevanov AV, Vishnyakova HS, Kireev II, Dunina-Barkovskaya AY (2009) Serum-induced inhibition of the phagocytic activity of cultured macrophages IC-21. Biochemistry (Mosc) Suppl Ser A Membr Cell Biol 3:412–419

18. Phanse Y, Ramer-Tait AE, Friend SL, Carrillo-Conde B, Lueth P, Oster CJ, Phillips GJ, Narasimhan B, Wannemuehler MJ, Bellaire BH (2012) Analyzing cellular internalization of nanoparticles and bacteria by multi-spectral imaging flow cytometry. J Vis Exp 64:e3884. https://doi.org/10.3791/3884

Part III

Imaging Flow Cytometry: FlowCam

Chapter 13

FlowCam 8400 and FlowCam Cyano Phytoplankton Classification and Viability Staining by Imaging Flow Cytometry

Kathryn H. Roache-Johnson and Nicole R. Stephens

Abstract

This chapter provides a protocol for a detailed evaluation of phytoplankton and nuisance cyanobacteria with the FlowCam 8400 and the FlowCam Cyano. The chapter includes (i) detailed description of the quality control of fluorescent mode of the FlowCam, (ii) detailing methods for discriminating nuisance cyanobacteria using the FlowCam Cyano, how to set up libraries and classification routines for commonly used classification reports, and (iii) detailing methods for viability staining to quantify LIVE versus DEAD phytoplankton using the FlowCam 8400.

Key words Chlorophyll fluorescence, Classification, FlowCam, Freshwater cyanobacteria, Imaging flow cytometry, Phycocyanin fluorescence, Phytoplankton, Viability staining

1 Introduction

The FlowCam® was developed at the Bigelow Laboratory for Ocean Sciences in East Boothbay, (ME, USA), and was first introduced in 1999. The name FlowCam is derived from 'flow cytometer and microscope'. The instrument was originally designed for the oceanographic community to image and count phytoplankton [1–3]. Since then, aquatic applications have expanded to include the use of FlowCam data and associated images to ground truth satellite phytoplankton remote sensing [4, 5], to identify and quantify harmful algal bloom (HAB) organisms, to forecast HAB events [6], to quantify and classify plankton community composition in environmental samples [2, 3], and for use by municipal drinking water companies to monitor for 'taste and odor' algae and other nuisance particles. A number of studies evaluating the FlowCam in comparison to light microscopy have found a strong, positive correlation between the two methodologies [7–12]. Additionally, FlowCam applications have expanded beyond aquatic

Natasha S. Barteneva and Ivan A. Vorobjev (eds.), *Spectral and Imaging Cytometry: Methods and Protocols*, Methods in Molecular Biology, vol. 2635, https://doi.org/10.1007/978-1-0716-3020-4_13,

studies, and it is now commonly used in the biopharmaceutical industry to identify protein aggregates, silicone oil, and other contaminants and improve the quality and safety of drug formulations [1, 13, 14].

Some notable improvements to FlowCam's original design include cross-polarizing lenses for evaluating the birefringence of veliger shellfish in a solution [15–17], oil immersion [18, 19] to increase the optical resolution to accommodate a 40× objective lens to image particles less than 1 micron (FlowCam Nano [40×]), a high-speed digital camera to accommodate up to 60 frames per second (FlowCam 8000 series [3]), low magnification to image zooplankton (FlowCam Macro [2, 3]), a circuit board to accommodate the camera's signaling to the electronics and software, a liquid handler robotic autosampler (FlowCam ALH [20]), a light obscuration module (FlowCam LO [21]), and software improvements that include a database structure for analyzing multiple runs at one time (VisualSpreadsheet™ 5).

Most FlowCam models use a syringe pump to accurately measure the fluid volume analyzed and can be configured to include various objective sizes to increase the magnification of the image to the camera sensor. A variety of flow cells act as cuvettes for the sample to pass through, and, like a microscope slide, there are different focal planes that exist within the flow cell. The sample passes in front of the magnifying objective, and the flashing LED acts like a strobe light to suspend the particles in motion so that the digital camera can take pictures as the software evaluates the background from the dark and/or light pixels (i.e., particles) and crops out the images into a collage window. Groups of pixels that represent particles are then "segmented" out of each raw image and saved as a separate collage image, along with more than 40 different morphological sizing and shape parameter measurements for each particle. VisualSpreadsheet can be used to run samples, analyze images, compare concentrations and particle parameters across runs, perform image analysis, and create automated classification routines to categorize phytoplankton and other particles.

Over the past 20 years, Yokogawa Fluid Imaging Technologies has made significant improvements to the design of the FlowCam, and in 2016 introduced the FlowCam 8000 series. In the FlowCam 8000 series, the model this chapter addresses, there are four objectives available: 2×, 4×, 10×, and 20×. Flow cells of high optical quality (fused silica) display the entire width of the flow cell in the field of view (FOV) of the camera. A different FOV flow cell is recommended for each objective depending on the size range of the particles being analyzed (see Table 1 for details). There are two LED options depending on whether the FlowCam is equipped with a black and white camera (blue LED for maximum resolution) or a color camera (white LED).

Table 1
Recommended particle cell size ranges (cell diameter), objective, syringe, and flow cell sizes for the FlowCam 8000 series (8400, Cyano) using trigger mode

Objective lens	FOV flow cell depth (μM)	Syringe size (mL)	Suggested flow rate (mL/min)	Minimum particle size (μM)	Maxium particle size (μM)
2×	1000	12.5	5.0	70	1000
4×	600	5	0.9	4	600
4×	300	5	0.9	4	300
10×	100	1	0.15	4	100
20×	50	0.5	0.03	3	50

Table 2
FlowCam 8400 parameters and targeted algal pigments and viability stains

Laser excitation wavelength	Optical filters Channel 1	Pigment target	Optical filters Channel 2	Stain, pigment
488 nM (blue)	650 longpass	Chlorophyll	Bandpass 525/30	FDA (fluoresceindiacetate), GRS (green fluorescein reactive stain)
532 nM (green)	650 longpass	Chlorophyll	Bandpass 575/30	RFS (red fluorescent reactive dye), phycoerythrin
633 nM (red)	Bandpass 700/10	Chlorophyll	650 bandpass 650/10	FarRed, phycocyanin

The FlowCam 8000 series includes the FlowCam 8100, which contains a high-speed digital camera that shutters at a customizable frame rate (AutoImage mode), and the FlowCam 8400, which is also equipped with a single laser-enabled excitation wavelength combined with two photomultiplier detectors (PMTs or channels, Ch) to trigger the camera (Trigger mode). In addition to triggering the camera, relative fluorescence information is gathered for each camera frame captured, with the caveat that the associated images are not epifluorescent representations. The laser options for the FlowCam 8400 include blue (488 nm), green (532 nm), or red (633 nm) lasers where one channel is tuned for chlorophyll and the second is suitable for a secondary emission range (Table 2).

The optical layout of the FlowCam 8000 series is depicted in Fig. 1. The dotted line in Fig. 1b indicates the additional laser and PMTs for the FlowCam 8400 (not included in the FlowCam 8100 model).

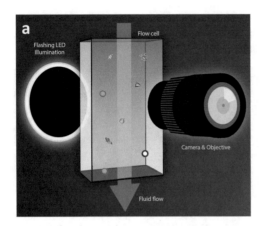
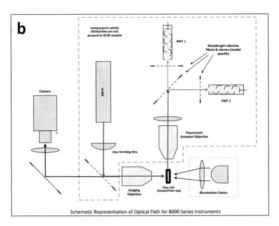

Fig. 1 (**a**) A schematic rendering how the FlowCam's camera captures particles that pass through a flow cell. (Image courtesy of Sarah Isakson); (**b**)The optical layout of the FlowCam 8000 series hardware

The FlowCam Cyano model is equipped with the 633 nm laser [22] as the two detectors are optimized for detecting chlorophyll and phycocyanin pigments that are present in freshwater cyanobacteria and used to characterize cyanobacterial blooms and discriminate cyanobacteria from other types of phytoplankton and detritus [23–25]. The Manual for real-time quality control of phytoplankton by NOAA (National Oceanic and Atmospheric Administration, USA) describes the methods for calibrating the FlowCam for sizing and counting microalgae as well as descriptions of other commonly used techniques for studying phytoplankton [26]. In addition, this manual outlines the current programs and subject matter experts in the phytoplankton monitoring community throughout the United States.

A comparison of different imaging flow cytometers for phytoplankton has already been evaluated [25], and multi-instrument assessment of phytoplankton for abundance and cell sizes has been used widely [27]. A thorough examination of FlowCam's imaging flow cytometry capabilities for quantification and classification of phytoplankton has also been written [25, 28]. This chapter addresses recommended approaches specifically for evaluating phytoplankton and nuisance cyanobacteria with the FlowCam 8400 [5] and the FlowCam Cyano [10] in greater detail.

Viability staining of microalgae has been used widely with FlowCams equipped with laser-enabled trigger mode [29–31]. Those studies were performed on previous FlowCam models (e.g., FlowCam VS-IV), and this chapter provides updated viability staining recommendations for use with the FlowCam 8400. This application can be used to assess the viability of cells during experiments with different stressors, such as pesticide, wastewater, or ballast water treatments [32–34].

2 Materials

2.1 Algal Cultures and Natural Samples Collection

Algal cultures used in this chapter included cyanobacterial and green algae cultures (*Anabaena, Cosmarium, Gleocapsa, Staurastrum*) from Carolina Biological Supply. The growth media used was Alga-Gro Freshwater medium (Carolina Biological Supply cat # 153752). Cultures were kept in a room temperature cabinet in continuous light with standard fluorescent bulbs.

Natural samples used as examples in the FlowCam Cyano fluorescent filter section were obtained from lakes in Colorado and Kansas during the summers of 2017 and 2018 using a 20 μm plankton net. The samples were run using trigger mode and diluted to Particle Per Used Image (PPUI) less than 1.5.

2.2 Instrumentation Accessories and Software

1. The FlowCam Cyano, equipped with a red excitation laser (633 nM), was used to discriminate cyanobacteria from other algae based on fluorescence emission data.

2. Microalgae viability staining: A FlowCam 8400 equipped with either a blue (488 nm), green (532 nm), or red (633 nm) laser was used for viability staining of live or dead microalgal cells. See Figs. 1 and 2 for the FlowCam instrumentation and optical layout diagrams of the components inside the covers.

   For 8400 FlowCam instruments, a 10× objective, flow cell with FOV (Field of view)100, and 1.0 mL syringe were installed. The standard Aquatic context settings were used (*see* **Note 1**). Samples were assessed prior to the experiments to ensure a PPUI of less than 1.5, and the "Used Image Percentage" was relatively high (above ~65%) (*see* **Notes 2** and **3**).

3. VisualSpreadsheet vs. 4 or 5 (Yokagawa Fluid Imaging Inc.) software was used to acquire all data runs, while VisualSpreadsheet vs. 5 was used to perform all data analysis.

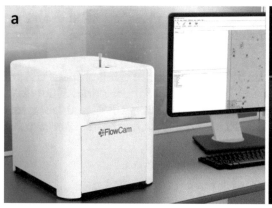
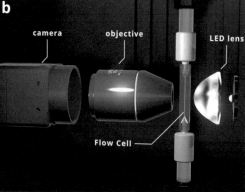

Fig. 2 FlowCam 8000 Series laboratory benchtop unit. (**a**) Exterior view of the 8000 series FlowCam; (**b**) Optical layout of the camera (objective, flow cell, and LED)

Table 3
Fluorescent beads from different manufacturers

Manufacturer	Brand name	Catalog #	Bead Diameter, μm	FlowCam 8400 laser, Excitation	Emission wavelength max, nm	Channel 1	Channel 2
ThermoFisher	Fluoromax red	36-4	13	488, 532	612	X	
ThermoFisher	Fluoromax green	35-4	15	488	508		X
Bangs	FITC	891	7–9	488	519		X
Bangs	R-Phycoerythrin	899	7–9	532	578		X
Spherotech	Sky blue	FP-15070-2	16.2	633	700	X	
Spherotech	Cy blue	FP-15066-2	15.7	633	650		X

4. *Fluorescent beads for FlowCam instrument quality control.* In flow cytometry, fluorescent beads are commonly used to confirm the performance of the PMT (photomultipliers) channels because, unlike phytoplankton, the fluorescence does not change as long as the suspension is protected from ambient light. Phytoplankton's natural autofluorescence can shift depending on light, growth conditions, or the health of the algal culture. We recommend using both fluorescent beads and healthy (late exponential growth or early stationary phase) algal cultures to assess and confirm your FlowCam 8400's laser and PMT channels performance.

Table 3 outlines a variety of fluorescent beads from different manufacturers that can be used to qualify the performance of the operation of the laser and PMTs in FlowCam 8400 s, and the appropriate PMT channel in which the beads will fluoresce optimally.

5. Viability stains. See Table 2 for a list of viability stains and the appropriate laser and PMT channel for detection. Fluorescein diacetate (FDA) preparation: store at −20 °C and laboratory grade DMSO was used as the solvent (*see* **Note 4** about using acetone as the solvent).

6. 15 mL centrifuge tubes for algal samples.

7. Polypropylene vials with screw caps for FDA staining.

3   Methods

3.1   How to Qualify the Laser and PMT Performance in FlowCam 8400S Using Fluorescent Beads

1. QC (Quality control) of FlowCam instrument equipped with 488 nm, CH1 (650LP), and CH2 (535/30) (Fig. 3a–c).

   The red population corresponds to Fluoromax Red beads, the green population corresponds to Fluoromax Green beads, and the purple population corresponds to Bangs FITC beads. Notice in the CH1 vs. CH2 Peak plot (a) that the FITC and Green beads are combined and separated from the Red beads. If we evaluate the CH2/CH1 ratio relative to Diameter ABD

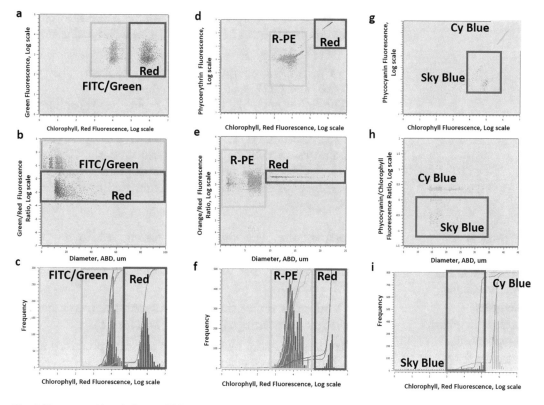

Fig. 3 Fluorescent beads for qualifying the FlowCam laser and PMT detector performance. Far left panels (**a–c**) display graphs from 488 nm (blue laser) FlowCam 8400 (with optical filters 650LP and 525/30 at CH1 and CH2). FITC-beads – green rectangle; Fluoromax Red beads – red rectangle. (**a**) dotplot of CH1 (Green fluorescence) vs. CH2 (chlorophyll fluorescence), (**b**) dotplot of Green: Red ratio fluorescence vs. diameter ABD (μm), (**c**) histogram plot of chlorophyll vs. red fluorescence. Middle panels (**d–f**) display graphs for FlowCam 8400 532 nm laser (with optical filters 650LP and 575/30 at CH1 and CH2) withR-Phycoerythrin-beads (Bangs Inc.) -orange rectangles and Fluoromax Red – red rectangles: (**d**) PE CH2 orange fluorescence vs. chlorophyll CH1 fluorescence, (**e**) orange:red CH2:CH1 ratio vs. diameter ABD (μm), and (**f**) histogram plot of CH1 chlorophyll red fluorescence. Right panels (**g–i**) display graphs for 633 nm FlowCam 8400 Cyano (with optical filters 700/10 and 650/10 at CH1 and CH2). The fluorescent beads used were Spherotech Cy Blue fluorescent beads -light blue rectangle and Sky Blue fluorescent beads -dark blue rectangle: (**g**) Phycocyanin CH2 fluorescence vs. Chlorophyll CH1 fluorescence, (**h**) Phycocyanin:Chlorophyll or CH2:CH1 ratio vs. diameter ABD (μM), and (**i**) histogram plot of chlorophyll CH1 fluorescence

(μm) (b) and Chlorophyll, Red fluorescence Intensity plot, we see that the Green and FITC beads have a higher signal than the red beads.

2. QC of FlowCam equipped with 532 nm laser and fluorescent CH1 (650LP) and CH2 (575/30) (Fig. 3d–f).

   The red population corresponds to the Fluoromax Red beads, while the orange population corresponds to the Bangs R-PE (Phycoerythrin) beads (Bangs Laboratories Inc.). The two types of beads are displayed as different populations in the CH1 peak vs. CH2 peak plot (**d**), the CH2/CH1 ratio vs. Diameter ABD (Area Based Diameter) (μm) plot (**e**), and the CH1 peak frequency histogram (f). We can see that the CH2 vs. CH1 scatterplot and chlorophyll, Red fluorescence histogram that the red beads display a higher signal compared to the R-PE beads.

3. QC of FlowCam equipped with 633 nm laser with CH1 (700/10) and CH2 (650/10) (Fig. 3g–i).

   The Spherotech Sky blue and Cy blue beads (Spherotech Inc., USA) display as separate populations in the CH1 vs. CH2 peak (**g**), CH2/CH1 Ratio vs. Diameter ABD plot (**h**), and CH1 fluorescence histogram (**i**). As expected, the CH2/CH1 ratio and chlorophyll fluorescence histogram display higher fluorescence for the Cy blue beads compared to the Sky Blue beads.

3.2 How to Distinguish Cyanobacteria with Phycocyanin and Chlorophyll Fluorescence from Phytoplankton Containing Only Chlorophyll in the FlowCam Cyano Instrument

1. To prepare the algal cultures for testing, add a few drops (~100 μL) of each culture to a clean 15 mL tube using a separate transfer pipet for each culture. Fill the tube to the top with dH_2O.

2. Run 1 mL of a sample using the instrument with recommended Aquatic context settings (Fig. 4 and *see* **Note 1**).

   1a. To prepare natural samples, net tows are preferred in the winter season or when there is low algal density.

   2a. Grab samples can be appropriate in the summer season when algal density is higher, but net tows are still preferred.

   3a. If cyanobacteria surface scum is present, a grab sample can be used. It is not practical to analyze scums with the FlowCam instrument for cell counting purposes, but scums should be noted when evaluating algal bloom presence. While the FlowCam instrument can be used to count cells, any method (including the microscope) has limitations for this type of sample. Significant dilution is required in both cases to optimize results.

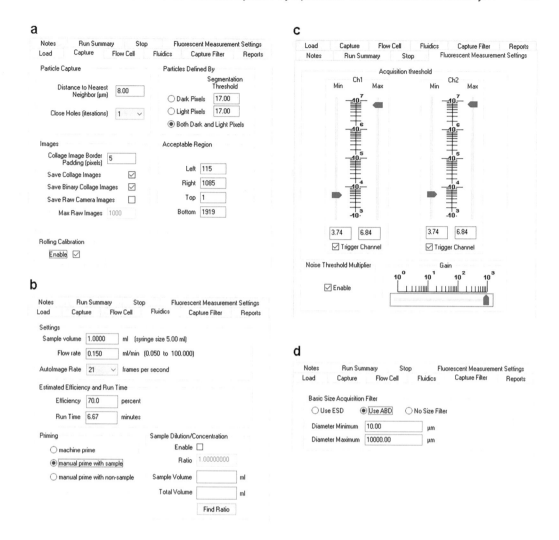

Fig. 4 Standard Aquatic Context settings for FlowCam 8400 instrument with 10× flow cell. Each tab of the context settings is displayed. (**a**) Capture settings indicate those associated with how the software parses out the images from the background; (**b**) Fluidics settings indicate how the sample is run through the flow cell by the syringe pump; (**c**) Fluorescence measurement settings indicate the sensitivity of the PMT detector channels; (**d**) The capture filter indicates the minimum diameter of acceptable particles that will be included in the collage view

3. After checking the laser alignment (Fig. 5), use the following scatterplots in VisualSpreadsheet software to assist in evaluating your data (Fig. 6). (1) Capture X vs. Capture Y (**a**, to view laser line and diagnose clogs), and Capture X vs. CH1 Peak (**b**, to view laser line and alignment), where the Capture X vs. Capture Y is the plot that displays all images in the physical layout of the FOV flow cell. Each dot corresponds to an image that was cropped out of the camera frame and placed into the collage window. The concentrated number of particles present represents the presence of the laser line, and its presence

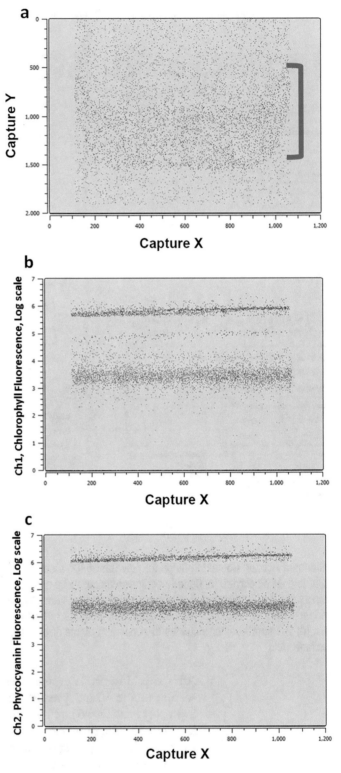

Fig. 5 The presence of a laser line confirms that your laser is properly aligned, and the laser lines can be viewed using appropriate fluorescent beads, cultures, or natural phytoplankton samples; (**a**) Capture X vs. Capture Y displays the actual physical space within your flow cell; (**b**) Capture X is the horizontal view of the flow cell vs. CH1 for chlorophyll; (**c**) CH2 for the second PMT detector

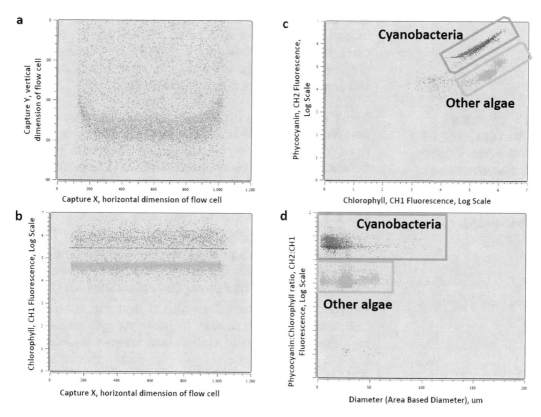

Fig. 6 FlowCam 8400 Cyano's instrument fluorescent signatures with 633 nm laser and associated detectors tuned for chlorophyll and phycocyanin fluorescence. Left panels display the physical dimensions inside the flow cell. Left upper panel (**a**) displays Capture X, the horizontal dimension of the FOV flow cell vs. Capture Y, the vertical dimension of the FOV flow cell, and left lower panel (**b**) displays the chlorophyll, CH1 fluorescence vs. Capture X horizontal dimension of FOV flow cell. The right panels display the separation of nuisance cyanobacteria and green algae using cultures from each group. The upper right panel (**c**) displays Phycocyanin CH2 fluorescence vs. Chlorophyll CH1 fluorescence, where the blue rectangle surrounds *Anabaena* and *Gloeocapsa,* and the bright green rectangle surrounds the *Cosmarium* and *Staurastrum*. The lower right panel (**d**) displays Phycocyanin: chlorophyll ratio where the upper blue rectangle surrounds the cyanobacteria, and the lower green rectangle surrounds the green algae. The green selected population is *Cosmarium* and *Staurastrum* combined as green dots and highlighted in the green rectangle

confirms that the laser is operational and in alignment (*see* **Note 5** for additional diagnostics for evaluating the laser and PMT performance in 8400 FlowCams). CH1 Peak vs. CH2 Peak (**c**, shows a separation of cyanobacteria vs. other algae), CH2/CH1 or Phycocyanin: Chlorophyll ratio vs. Diameter ABD (**d**, shows a separation of cyanobacteria vs. other algae).

4. We used cultures of cyanobacteria (*Anabaena* and *Gloeocapsa*) and green algae (*Cosmarium* and *Staurastrum*) to optimize the VisualSpreadsheet fluorescent filters (*see* **Note 6**).

5. Each of the cultures was diluted so that the particles per used image (PPUI) was near or less than 1.5 (*see* **Note 2**). Notice in these two plots that the two cyanobacteria (*Anabaena* and *Gloeocapsa*) group together while the green algae (*Cosmarium* and *Staurastrum*) group together (Fig. 7c, d).

3.3 Fluorescent Filters Can Discriminate Nuisance Cyanobacteria from Other Algae and Detritus in Natural Freshwater Samples

1. The FlowCam Cyano has three pre-built fluorescent filters installed. They are filters for cyanobacteria, diatoms and other algae, detritus, and decomposing particles. These pre-built fluorescent filters are based upon the CH2/CH1 ratio fluorescent values from microalgae that have been collected on a wide range of natural freshwater samples throughout North America and can be modified to be used with cultures or natural samples.

2. Natural algal samples can be viewed, and the different groups can be separated using fluorescent filters to help distinguish cyanobacteria from other phytoplankton and detritus (Fig. 7).

3. Libraries can be created from natural algal samples collected from a certain region to assist in image recognition and help with the classification of your natural samples into different reports. Both CH1 vs CH2 peak and Diameter ABD vs. CH2/CH1 ratio plots can be used to discriminate the cyanobacteria that contain phycocyanin from the other phytoplankton that do not contain phycocyanin (Fig. 7b, c).

4. VisualSpreadsheet software can be used to create new fluorescent filters and assist in automatically separating and calculating the count and concentration of different algal populations.

5. If the middle cyanobacteria population (blue population) is highlighted in the CH2/CH1 ratio vs. Diameter ABD plot (Fig. 7c), the same population will be highlighted in the CH1 vs. CH2 peak plot (Fig. 7b), and the corresponding images can be viewed in collage window and sorted by CH2/CH1 ratio values (Fig. 7a, d, e).

6. The minimum and maximum CH2/CH1 ratio values will appear in the particle property table and could be used to create or modify the pre-built fluorescent filter for cyanobacteria population.

7. Likewise, the lower fluorescent population of other algae can be highlighted (Fig. 7b, c) so that the corresponding fluorescent filter can be created.

8. Once these fluorescent filters are created with algal cultures, they can be slightly modified using natural samples (Fig. 8) so that the cyanobacteria can be separated from other algae and detritus or decomposing algae.

9. By viewing the images based upon CH2/CH1 ratio and sorting by greatest to least, the maximum value can be confirmed to be correct, and the associated images, contain mostly cyanobacteria or other phycocyanin containing algal genera.

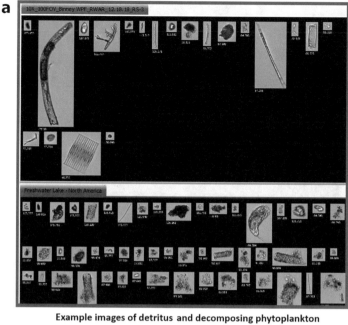

a

Example images of detritus and decomposing phytoplankton

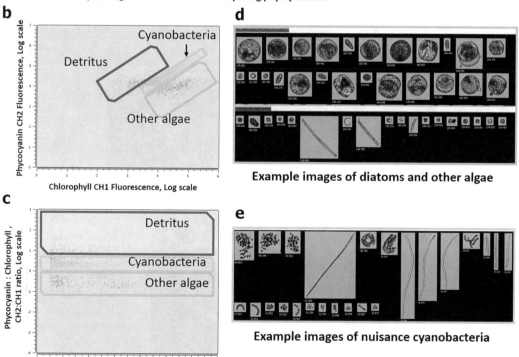

b

Cyanobacteria

Detritus

Other algae

Chlorophyll CH1 Fluorescence, Log scale

Phycocyanin CH2 Fluorescence, Log scale

c

Detritus

Cyanobacteria

Other algae

Phycocyanin : Chlorophyll , CH2:CH1 ratio, Log scale

Diameter, Area Based Diameter, um

d

Example images of diatoms and other algae

e

Example images of nuisance cyanobacteria

Fig. 7 North American freshwater natural samples taken during the summer season. The data was acquired with FlowCam 8400 Cyano. (**a, d, e**) FlowCam Cyano's signature plots of freshwater natural samples showing the separation of cyanobacteria containing phycocyanin from other algae and detritus/decomposing phytoplankton; (**b**) phycocyanin compared to chlorophyll fluorescence; (**c**) Diameter ABD (μm) vs. phycocyanin CH2: chlorophyll CH1 ratio. The selected images and populations in the graphs are highlighted in red for detritus and decomposing phytoplankton, light blue for cyanobacteria, and green for diatoms and other algae that do not contain phycocyanin pigment

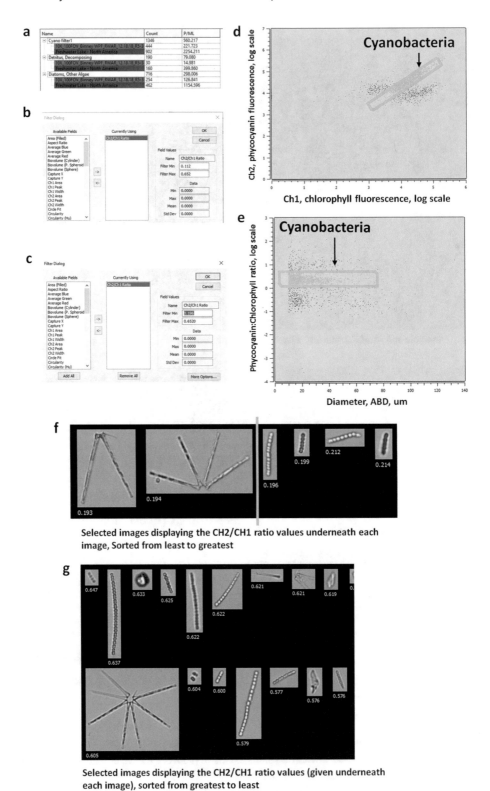

Selected images displaying the CH2/CH1 ratio values underneath each image, Sorted from least to greatest

Selected images displaying the CH2/CH1 ratio values (given underneath each image), sorted from greatest to least

Fig. 8 Fluorescent filters using CH2 phycocyanin:CH1 chlorophyll fluorescence log ratio to separate nuisance cyanobacteria from other algae that do not contain phycocyanin. Far left panels display an example of the filter table with the pre-build fluorescent filters and two sample runs (**a**) and VisualSpreadsheet filter windows used

3.4 Discrimination of Algal Types by Visual Determination of Filter Boundary Using the "View Window"

The "View Window" can be used to visually determine the filter boundary and discriminate cyanobacteria (Fig. 9). This method works well when viewing the images and visually distinguishing cyanobacteria from green algae and diatoms.

1. Open a run that contains a good mix of cyanobacteria and other algae/diatoms and sort images based on the CH2/CH1 ratio. The values for CH2/CH1 ratio will appear underneath each image.

2. Sort the particles from the lowest CH2/CH1 Ratio (in this case, algae and diatoms) to the highest CH2/CH1 Ratio (in this case, cyanobacteria).

3. Scroll through the images until a shift from one category to the other occurs (Fig. 9c). It is rare that there is a distinct switch from one type to the other, and there may be a minor carryover. In Fig. 9c, the image subset shows the shift from algae and diatoms to cyanobacteria, and it is marked by an orange line.

4. The value displayed can be entered as the minimum value in the fluorescent filter for cyanobacteria.

3.5 Discrimination of Algal Types by Using the Ch2:Ch1 log ratio-Based Filter

1. The second method uses the CH2/CH1 or phycocyanin to chlorophyll log ratio vs. diameter (ABD) scatterplot to graphically determine the filter boundary. This method allows you to assign filter values based upon the separation of the categories on the scatterplot of CH2/CH1 Log ratio vs. diameter. The three categories will separate into detritus and decomposing, cyanobacteria, diatoms, and other algae.

2. The Diatoms/Other algae section of the plot can be delineated with the boundary of the Cyanobacteria section of the plot.

3. The associated images will be displayed in the View Window and they can be sorted by the CH2/CH1 ratio values.

4. In the same manner as the first method, the boundary CH2/CH1 ratio values can be found and entered as the fluorescent filter values.

Fig. 8 (continued) to create or edit filter values: filter window before evaluating the images (**b**) and after evaluating the images and modifying the minimum filter value (**c**). Center panels display the signature plots that enable the discrimination of cyanobacteria from other phytoplankton.Top graph (**d**) is CH2 phycocyanin fluorescence compared to CH1 chlorophyll fluorescence, and lower graph (**e**) is the CH2:CH1 phycocyanin: chlorophyll ratio compared to the diameter ABD (μm). The right panels display selected images using the fluorescent filter for cyanobacteria based upon CH2/CH1 ratio. Upper image (**f**) displays the CH2/CH1 ratio values sorted from least to greatest. After examination of the images with the CH2/CH1 ratio values displayed underneath each image, the minimum value needs to be adjusted from 0.112 to 0.196 (blue line), where the cyanobacteria begin to appear (**f**). The lower images (**g**) are the CH2/CH1 ratio values sorted from greatest to least and reveal that the CH2/CH1 ratio maximum value is set correctly

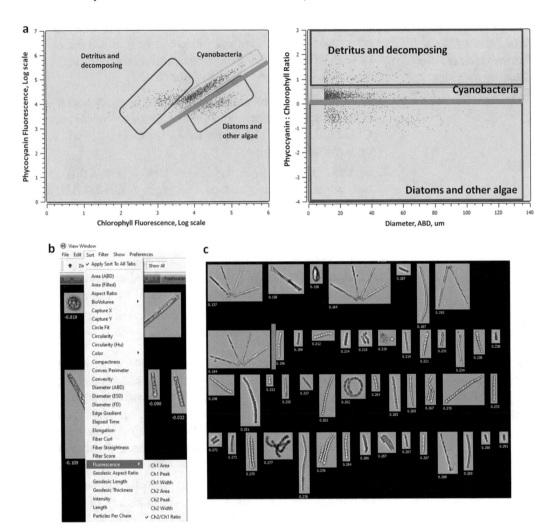

Fig. 9 How to use the view window to visually determine the filter boundary using the phycocyanin: chlorophyll fluorescence for a typical mixed phytoplankton sample that contains cyanobacteria and other algae (**a**). How to view the CH2/CH1 ratio values in VisualSpreadsheet vs. 5 (**b**). In VisualSpreadsheet vs. 5, images can be sorted to display the CH2/CH1 ratio; values underneath each color digital image and the orange line indicates the shift from algae to cyanobacteria (**c**)

5. If there is a mismatch of desired cyanobacteria, then the delineated region can be redrawn. Cyanobacteria begin to appear when the CH2/CH1 ratio value is near 0.196 (Fig. 9c).

6. Most particles to the right of the highlighted particle are cyanobacteria, so 0.196 would be a good value to use for discriminating cyanobacteria from diatoms and green algae.

3.6 Using Libraries to Help Sort Algal Sample Runs

The Visualspreadsheet software contains more than 25 pre-built freshwater libraries (at the time of this writing – on-site currently available are only libraries for *Anabaena, Aphanazomenon, Asterionella, Cylindrospermopsis, Fragilaria, Lyngbya, Microcystis, Pediastrum, Planktothrix, Rotifers, Scenedesmus*) from samples across North America that represent the dominant nuisance organisms for water utilities and monitoring agencies. In addition, there are more than 25 marine phytoplankton libraries available (mostly from the Gulf of Maine, USA). Some of these libraries will need to be optimized but can serve as a visual tool to build libraries from different regions. Due to the diverse morphology exhibited by planktonic organisms and the variety of context settings used, VisualSpreadsheet may not be able to correlate the pre-built libraries with samples (*see* **Notes 7–8** for library considerations).

3.7 Viability Staining with GRS Dye

1. For algal sample preparation, two aliquots of 10 mL of each culture were prepared in 15 mL centrifuge tubes and diluted so that the PPUI was less than 1.5 and then stained.

2. For the "Live Test", an aliquot was stained immediately after dilution. For the "Dead Test", an aliquot was stained after incubation in a water bath heated to 70–90 °C for 20–30 min. A thermometer placed in a 15 mL tube filled with dH_2O was used as a proxy for the temperature inside the "Dead Test" tubes. All staining incubation steps were carried out in the dark.

3. For Green Reactive Stain (GRS), EX/EM 495/520 nM) preparation, thawed out and stored at −20 °C to room temperature.

4. Add 50 μL DMSO, Component B, to Component A (reactive stain) and use within a few hours. The rest of the solution can be stored for up to 2 weeks at −20 °C. This is the working stock.

5. Add 1 μL of the working stock to 1 mL of sample and incubate at room temperature in the dark for at least 30 min. Figure 10 displays the results of viability staining with GRS dead and live cells from *Cosmarium* and *Staurastrum* algal cultures. The FlowCam instrument was equipped with the blue laser (488 nm), and the GRS-stained cells appeared in CH2 (bandpass 525/30).

3.8 Viability Staining with Fluorescein Diacetate Dye

1. An initial primary stock solution of Fluorescein diacetate (FDA, EX/EM 490/526 nm) was prepared by mixing with DMSO into a 5 mg/mL solution.

2. A working stock was prepared using 25 μL FDA primary stock solution into 10 mL dH_2O and stored at 4 °C.

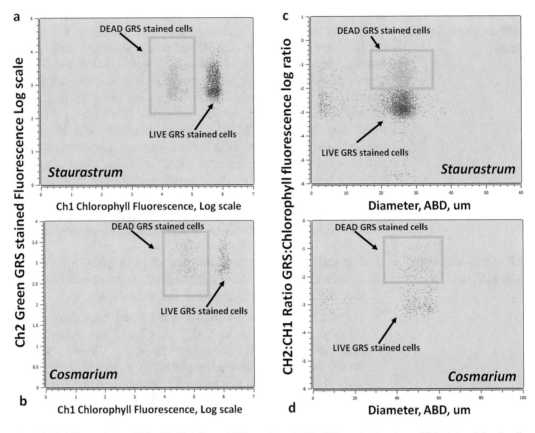

Fig. 10 Viability staining GRS with FlowCam 8400 equipped with 488 nm laser where GRS stained dead cells should take up more fluorescent stain than GRS stained live cells. (**a**, **c**): *Staurastrum*, dead GRS stained cells appear as the green population in CH2 compared to CH1 chlorophyll fluorescence (**a**) and graphs (**c**) CH1:CH2 Green: chlorophyll fluorescence ratio compared to diameter ABD (μm);(**b**, **d**): *Cosmarium*, dead GRS stained cells appear as the green population in CH2 Green fluorescence compared to CH1 chlorophyll fluorescence (**b**) and CH1:CH2 Green: chlorophyll fluorescence ratio compared to diameter ABD (μm) (**d**)

3. An aliquot of 250 μL of working stock was added to a 1 mL sample and incubated for at least 10 min in the dark at room temperature.

4. Figure 11 shows differences in fluorescence intensity of *Staurastrum*, *Cosmarium*, *Anabaena*, and *Gloeocapsa* live cells stained with the FDA from the dead algal FDA-stained cells.

3.9 Viability Staining with RFS Dye

1. Red Fluorescent Stain (RFS, EX/EM 595/615 nm), fluorescent dye preparation: Thaw out to room temperature.

2. Add 50 μL DMSO, Component B, to Component A (reactive stain) and use it within a few hours. The rest of the solution can be stored for up to 2 weeks at −20 °C. This is the working stock.

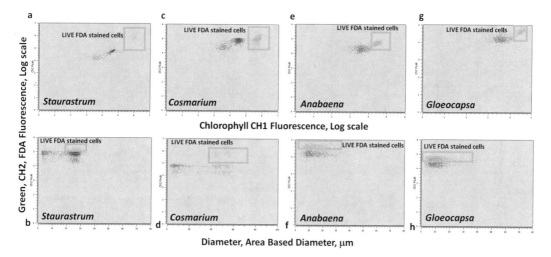

Fig. 11 Viability staining FDA FlowCam 8400 equipped with 488 nm laser(**a**, **b**) *Staurastrum*, FDA stained cells highlighted in bright green, (**c**, **d**) *Cosmarium*, FDA stained cells highlighted in bright green, (**e**, **f**) *Anabaena*, FDA stained cells highlighted in bright green, and (**g**, **h**) *Gloeocapsa*, FDA stained cells highlighted in bright green

3. Add 1 μL of the working stock to 1 mL of the sample. Let the sample incubate at RT in the dark for at least 30 min. Figure 12 demonstrates viability staining of *Cosmarium, Staurastrum, Anabaena,* and *Gloeocapsa* with RFS dye displaying a different population from LIVE RFS stained cells. The FlowCam 8400 was equipped with a green laser (532 nm).

3.10 Viability Staining with FarRFS Dye

1. Far Red Reactive Stain (FarRFS, EX/EM 650/665 nm), fluorescent dye preparation: Thaw out to room temperature.

2. Add 50 μL DMSO, Component B, to Component A (reactive stain) and use within a few hours. The rest of the solution can be stored up to 2 weeks at −20 °C. This is the working stock.

3. Add 1 μL of the working stock to 1 mL of the sample.

4. Let the sample incubate at RT in the dark for at least 30 min. Figure 13 displays the viability staining with FarRed in *Cosmarium* and *Staurastrum* where the dead FarRed stained cells showed a clearly different population from the live FarRed stained cells. However, the two cyanobacteria cultures tested, *Gloeocapsa* and *Anabaena*, were unable to differentiate the dead FarRed stained cells apart from the live FarRed stained cells. This is likely due to the interference of the phycocyanin that is inherently present in these two cyanobacteria. Therefore, this staining procedure should not be used when assessing mixed phytoplankton populations that may contain phytoplankton with phycocyanin. This protocol appears to be useful only for phytoplankton not containing phycocyanin pigment.

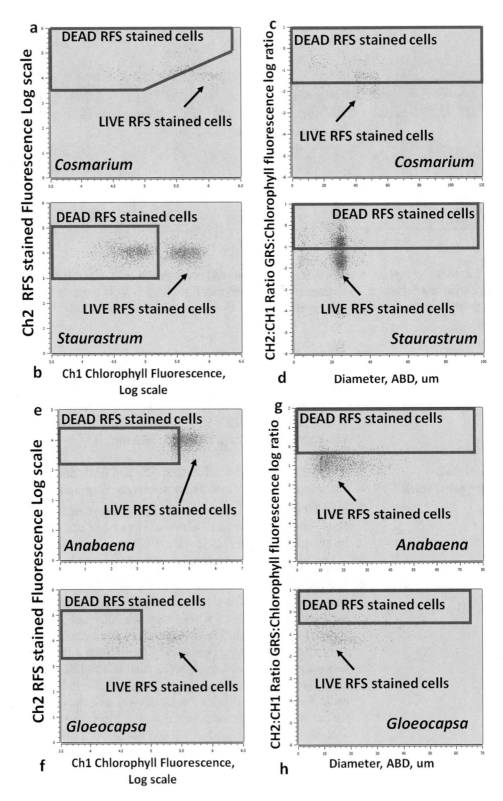

Fig. 12 Viability staining RFS with FlowCam 8400 instrument equipped with 532 nm laser where RFS stained dead cells should take up more fluorescent stain than RFS stained live cells. Red population is RFS stained dead cells while blue population is RFS stained live cells. (**a, c**) *Cosmarium* dead RFS-stained cells appear as

4 Notes

1. Standard Aquatics context settings for Trigger mode displaying Capture, Fluidics, Fluorescent Measurement Settings, and Capture Filter within your Context settings for data acquisition (see Fig. 4). These are the recommended context settings to use on your FlowCam instrument.

2. Particles Per Used Image (PPUI) refers to how many particles are in the image frame (on average) every time the laser is triggered. If more than one particle is in the flow cell when a fluorescent event triggers the camera's shutter release, all particles will be assigned the same fluorescence value, making it difficult to know which particle corresponds with the true fluorescence reading. This is problematic when using filters based on fluorescence values of cyanobacteria vs. diatoms and other algae to automatically classify particles by type or when discriminating between live vs. dead cells. When PPUI is too high, the incidence of misclassification due to incorrect fluorescence values being assigned increases, resulting in more manual correction for the operator. When running in trigger mode, the operator should verify that the PPUI is 1.0–1.5. The running PPUI value can be seen during image capture in the capture window at the bottom left-hand side of the screen. Run the sample using Trigger Mode and make a note of the PPUI. If the PPUI is less than 1.5, then no dilution is necessary. If the PPUI is greater than 1.5, estimate the dilution needed, assuming a linear relationship, and then re-run the sample using Trigger Mode again to confirm that your PPUI is 1.5 or less; if not, dilute further. After a Trigger mode data run completes, the PPUI for each run can be viewed in the Run Summary section of the Context table.

3. The "Used Image Percentage" is the ratio of the "Used Image Count" to the "Total Image Count" The "*Used Image Count*" refers to the number of image frames where at least one particle was segmented and captured (from Visualspreadsheet

---

Fig. 12 (continued) the Red population in CH2; RFS fluorescence compared to CH1 chlorophyll fluorescence (**a**), and CH1:CH2 RFS:chlorophyll fluorescence ratio compared to diameter ABD (μm) (**c**). (**b, d**) *Staurastrum* dead RFS stained cells appear as the red population in CH2; RFS fluorescence compared to Ch1 chlorophyll fluorescence (**b**) and Ch1:Ch2 RFS:Chlorophyll fluorescence ratio compared to diameter ABD (μm) (**d**). (**e, g**) *Anabaena* dead RFS-stained cells appear as the Red population in CH2 RFS fluorescence compared to CH1 chlorophyll fluorescence (**e**), and CH1:CH2 RFS:chlorophyll fluorescence ratio compared to diameter ABD (μm) (**g**). (**f, h**) *Gloeocapsa* dead RFS stained cells appear as the red population in CH2 RFS fluorescence compared to Ch1 chlorophyll fluorescence (**f**) and CH1:CH2, RFS:chlorophyll fluorescence ratio compared to diameter ABD (μm) (**h**)

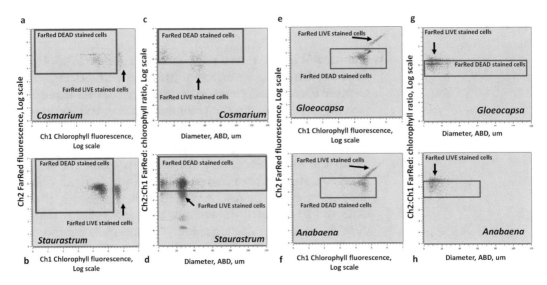

Fig. 13 Viability staining FarRed acquired with FlowCam 8400 Cyano instrument equipped with 633 nm laser. (**a, c**) *Cosmarium* dead FarRed stained cells appear as the Red population in CH2 FarRed fluorescence compared to CH1 chlorophyll fluorescence (**a**) and CH1:CH2, FarRed: chlorophyll fluorescence ratio compared to diameter (μm) (**c**). (**b, d**) *Staurastrum* dead FarRed stained cells appear as the red population in CH2 FarRed fluorescence compared to CH1 chlorophyll fluorescence (**b**), and CH1:Ch2, FarRed: chlorophyll fluorescence ratio compared to diameter ABD (μm) (**d**). e,g *Gloeocapsa* dead FarRed stained cells appear as the Red population in CH2 FarRed fluorescence compared to CH1 chlorophyll fluorescence (**e**) and CH1:CH2, FarRed: chlorophyll fluorescence ratio compared to diameter ABD (μm) (**g**). (**f, h**) *Anabaena* dead FarRed stained cells appear as the red population in CH2 FarRed fluorescence compared to CH1 chlorophyll fluorescence (**f**) and CH1:CH2, FarRed: chlorophyll fluorescence ratio compared to diameter ABD (μm) (**h**)

software). The "*Total Image Count*" is the total number of raw camera frames imaged, regardless of whether they contain a particle or not. By evaluating the "Used Image Percentage," a user can determine if the minimum fluorescence thresholds are set correctly. When using Trigger mode, the "Used Image Percentage" should be relatively high (above ~65%). Higher values indicate that when the laser is triggering an image capture event, it is picking up on actual particles and not background noise. If the minimum threshold value is too low, the laser will pick up on more background noise than particles resulting in a low "Used Image Count" relative to the "Total Image Count". You can check your "Used Image Percentage" for a run in the Run Summary section of the Context Summary. A low Used Image Percentage can occur if there are particles smaller than the minimum diameter set in the filter tab within context, decreasing this value may increase your Used Image Percentage.

4. We evaluated the use of acetone with FDA staining and found that it did not work for *Anabaena*, *Cosmarium*, or *Gloeocapsa*;

however, it did appear to differentiate between dead and live replicates of *Staurastrum*.

5. Evaluating Graphs to determine the laser performance (Fig. 6). Figure 6 displays cultures and fluorescent beads that were run on the FlowCam Cyano (FlowCam 8400 with 633 nm laser). The same phenomena exist for all types of 8400 FlowCam instruments. The graph Capture X and Y (Fig. 6a) demonstrates the presence of a laser line (inverted parabola, bracket). The laser line is where a concentrated number of images are captured when a fluorescent particle passes the laser beam, is excited, and emits fluorescence to the PMTs (channel 1 and channel 2) that is above the thresholds set in the fluorescent measurement within context settings (Fig. 4c). The absence of this laser line indicates a possible issue with the laser alignment. In addition to the Capture X and Y graph, there are additional graphs that display the tell-tale laser line to confirm the laser is operating correctly. Capture X vs. CH1 peak and Capture X vs. CH2 peak display the laser lines for both cultures and fluorescent beads. The absence of these diagnostic laser lines equals poor laser performance or misalignment.

6. *Optimizing fluorescent filter correlation to the CH1 (chlorophyll) vs. CH2 (phycocyanine) peak scatterplot.* The three fluorescence filters installed on your main window (Detritus/Decomposing, Cyanobacteria, and Diatoms/Other Algae) each correspond to a specific area on the CH1 Peak vs. CH2 Peak Scatterplot. As shown in Fig. 8, cyanobacteria make up a spike at ~45° angle in the relative center of the plot. Detritus and decomposing particles form a cluster that is above and left of the cyanine spike, while diatoms and other algae form a cluster below and to the right of the spike. Depending on your sample and the precise configuration of the PMT channel detectors of your instrument, the plot may vary somewhat, but the relative positions of each of the three groups will not change. The same fluorescence relative populations appear in Trigger mode using the 4× objective and flow cells FOV 300 or FOV600 and 20× objective and flow cell FOV50.

7. *Library considerations.* The number of libraries you build will depend on how many individual taxa you want to classify and report on. You will need a separate library for each of these taxa, but you may also need more than one library to capture a single taxon. If an organism has a considerable size or shape range, it may require 2–3 libraries to accurately represent its most common forms. An example of this is the cyanobacteria *Anabaena* spp. which may have coiled or straight chains. In this case, a separate library should be created for each morphological type. The more particle images a library contains, the better it accounts for the natural variability observed within a particular

group as long as the particles included are sufficiently uniform. It will take some time to build robust libraries for all taxa of interest, so a good starting point is 40–60 images per library. You can continue to add to libraries over time to increase the effectiveness of the statistical filters that are used to build classifications. The mantra for building libraries is to make them as "unique yet uniform" as possible. The more *unique* one group of organisms is compared to another, the more likely the FlowCam will be able to differentiate an image from other organisms. Remember that the software uses particle size and morphology to create statistical filters, so the more distinct a taxon is, the more its parameters will differ from other taxa. The collages in Fig. 9b represent taxa that are very distinctive and do not closely resemble other common plankton types. Taxa with similar sizes and morphologies will be more difficult to distinguish using statistical filters. The images in Fig. 9c show different taxa with similar sizes and morphological characteristics. Since these taxa are less unique, it will be more difficult to automate their classification. This does not mean an operator should not bother building libraries for these taxa, just that there will be more manual intervention required to classify them correctly. The more *uniform* the images are within a library, the more effective its statistical filter will be, as the statistical spread across the particle properties will be restricted. It may be necessary to separate different morphotypes or size ranges of a particular organism into different libraries to preserve uniformity, and it may be easier to accomplish uniformity for some taxa than others. Figure 7d is a partial library of *Cyclotella* which shows little variation in size or shape and is quite uniform. In contrast, the images of *Microcystis* exhibit a non-uniform, wide-ranging variety in size and morphology. A few library best practices include:

- Use images that are in reasonable focus (minimum edge gradient = 100).

- Do not use images that are cropped or cut off.

- Separate different morphotypes within the same taxon.

- Create separate libraries for each objective (e.g., 10×, 4×), data acquisition mode (i.e., AutoImage vs. Trigger Mode), and preservative used.

8. Using trigger mode, the categories can be broad (cyanobacteria, green algae/diatoms, or detritus) or specific (genus/species level). A simple classification may contain only a few categories while a more complex one might have 10 or more categories. The classification(s) you create will form the basis of your reporting strategy. Classifications can be semi-automated by incorporating statistical and/or value filters into class

definitions. Once you have created a classification and defined your classes and associated filters, it can be saved as a classification template. This template can then be applied to future data runs without needing to recreate the classification from scratch.

Acknowledgements

Support for this work was provided by Yokogawa Fluid Imaging Technologies. The authors would like to thank Frances Buerkens, Savannah Judge, Sarah Isakson, and Harry Nelson from Yokogawa Fluid Imaging Technologies for providing careful review of the manuscript.

References

1. Kiyoshi M, Shibata H, Harazono A et al (2019) Collaborative study for analysis of subvisible particles using flow imaging and light obscuration: experiences in japanese biopharmaceutical consortium. J Pharm Sci 108:832–841. https://doi.org/10.1016/j.xphs.2018.08.006

2. Krafft BA, Bakkeplass KG, Berge T et al. (2019) Report from a krill focused survey with RV Kronprins Haakon and land-based predator work in Antarctica during 2018/2019. Havforskningsinstituttet 2019. 103 p. Rapport fra havforskningen (2019-21). https://hdl.handle.net/10037/20019

3. Park J, Kim Y, Kim M, Lee WH (2018) A novel method for cell counting of *Microcystis* colonies in water resources using a digital imaging flow cytometer and microscope. Environ Eng Res 24:397–403. https://doi.org/10.4491/eer.2018.266

4. Blaschko MB, Holness G, Mattar MA et al (2005) Automatic in situ identification of plankton. In: Proceedings of the seventh IEEE Workshops on Application of Computer Vision (WACV/MOTION'05) – volume 1. IEEE Computer Society, USA, pp 79–86

5. Cetinić I, Poulton N, Slade WH (2016) Characterizing the phytoplankton soup: pump and plumbing effects on the particle assemblage in underway optical seawater systems. Opt Express 24:20703–20715. https://doi.org/10.1364/OE.24.020703

6. Buskey EJ, Hyatt CJ (2006) Use of the Flow-CAM for semi-automated recognition and enumeration of red tide cells (*Karenia brevis*) in natural plankton samples. Harmful Algae 5:685–692. https://doi.org/10.1016/j.hal.2006.02.003

7. Álvarez E, López-Urrutia Á, Nogueira E, Fraga S (2011) How to effectively sample the plankton size spectrum? A case study using Flow-CAM. J Plankton Res 33:1119–1133. https://doi.org/10.1093/plankt/fbr012

8. Álvarez E, Moyano M, López-Urrutia Á, Nogueira E, Scharek R (2014) Routine determination of plankton community composition and size structure: a comparison between FlowCAM and light microscopy. J Plankton Res 36:170–184. https://doi.org/10.1093/plankt/fbt069

9. Detmer TM, Broadway KJ, Potter CG, Collins SF, Parkos JJ, Wahl DH (2019) Comparison of microscopy to a semi-automated method (FlowCAM®) for characterization of individual-, population-, and community-level measurements of zooplankton. Hydrobiologia 838:99–110. https://doi.org/10.1007/s10750-019-03980-w

10. Graham MD, Cook J, Graydon J, Kinniburgh D, Nelson H, Pilieci S, Vinebrooke RD (2018) High-resolution imaging particle analysis of freshwater cyanobacterial blooms. Limnol Oceanogr Methods 16:669–679. https://doi.org/10.1002/lom3.10274

11. Hrycik AR, Shambaugh A, Stockwell JD (2019) Comparison of FlowCAM and microscope biovolume measurements for a diverse freshwater phytoplankton community. J Plankton Res 41:849–864. https://doi.org/10.1093/plankt/fbz056

12. Lehman PW, Kurobe T, Lesmeister S, Baxa D, Tung A, Teh SJ (2017) Impacts of the 2014 severe drought on the *Microcystis* bloom in San

Francisco Estuary. Harmful Algae 63:94–108. https://doi.org/10.1016/j.hal.2017.01.011

13. Singh R, Waxman L (2020) A streamlined bioanalytical approach to select a compatible primary container system early in drug development: a toolbox for the biopharmaceutical industry. J Pharm Sci 109:206–210. https://doi.org/10.1016/j.xphs.2019.09.016

14. Vargas SK, Eskafi A, Carter E, Ciaccio N (2020) A comparison of background membrane imaging versus flow technologies for subvisible particle analysis of biologics. Int J Pharm 578:119072. https://doi.org/10.1016/j.ijpharm.2020.119072

15. Spaulding BW (2009) Early detection can help eradicate invasive mussels. J AWWA 101:19–20. https://doi.org/10.1002/j.1551-8833.2009.tb09977.x

16. Yokogawa Fluid Imaging Technologies, Inc (2013) System and method for monitoring birefringent particles in a fluid. US Patent 8,345,239, 01 B1 January 2013

17. Yokogawa Fluid Imaging Technologies, Inc (2015) System and method for monitoring birefringent particles in a fluid. US Patent 9,151,943 B2, 06 October 2015

18. Sieracki C (2018) Extending the limits: oil immersion flow microscopy. http://www.labcompare.com/10-Featured-Articles/349589-Extending-the-Limits-Oil-Immersion-Flow-Microscopy/. Accessed 22 Jun 2020

19. Yokogawa Fluid Imaging Technologies, Inc (2018) Oil-immersion enhanced imaging flow cytometer. US Patent 2009/0273774 A1 Nov 5 2009

20. Graham G, Camp R (2017) Instrumentation advance speeds plankton study. Sea Technol 58:30–32

21. Detmer TM, Broadway KJ, Potter CG et al (2019) Comparison of microscopy to a semi-automated method (FlowCAM®) for characterization of individual-, population-, and community-level measurements of zooplankton. Hydrobiologia 838:99–110. https://doi.org/10.1007/s10750-019-03980-w

22. Yokogawa Fluid Imaging Technologies, Inc System and method for light obscuration enhanced imaging flow cytometry. US Patent 10,761,007 May 31, 2018

23. Yokogawa Fluid Imaging Technologies, Inc (2018) System and method for monitoring particles in a fluid using ratiometric cytometry. US Patent 9,983,115, Sept 21, 2015

24. Adams H, Buerkens F, Cottrell A, Reeder S, Southard M (2018) Use an integrated approach to monitor algal blooms. Opflow 44:20–21. https://doi.org/10.1002/opfl.1113

25. Lehman PW, Kurobe T, Teh SJ (2020) Impact of extreme wet and dry years on the persistence of *Microcystis* harmful algal blooms in San Francisco Estuary. Quat Int 521:16–25. https://doi.org/10.1016/j.quaint.2019.12.003

26. Dashkova V, Malashenkov D, Poulton N, Vorobjev I, Barteneva NS (2017) Imaging flow cytometry for phytoplankton analysis. Methods 112:188–200. https://doi.org/10.1016/j.ymeth.2016.05.007

27. Menden-Deuer S, Morison F, Montalbano AL et al (2020) Multi-instrument assessment of phytoplankton abundance and cell sizes in mono-specific laboratory cultures and whole plankton community composition in the North Atlantic. Front Mar Sci 7. https://doi.org/10.3389/fmars.2020.00254

28. Poulton NJ (2016) FlowCam: quantification and classification of phytoplankton by imaging flow cytometry. In: Barteneva NS, Vorobjev IA (eds) Imaging flow cytometry: methods and protocols. Springer, New York, pp 237–247

29. Gancel HN, Carmichael RH, Park K, Krause JW, Rikard S (2019) Field mark-recapture of calcein-stained larval oysters (*Crassostrea virginica*) in a freshwater-dominated estuary. Estuar Coasts 42:1558–1569. https://doi.org/10.1007/s12237-019-00582-6

30. Natunen K, Seppälä J, Schwenk D et al (2015) Nile Red staining of phytoplankton neutral lipids: species-specific fluorescence kinetics in various solvents. J Appl Phycol 27:1161–1168. https://doi.org/10.1007/s10811-014-0404-5

31. Shuman TR, Mason G, Reeve D et al (2016) Low-energy input continuous flow rapid pre-concentration of microalgae through electro-coagulation–flocculation. Chem Eng J 297:97–105. https://doi.org/10.1016/j.cej.2016.03.128

32. Fontvieille DA, Outagourouine A, Thevenot DR (1992) Fluorescein diacetate hydrolysis as a measure of microbial activity in aquatic systems: application to activated sludges. Environ Technol 13:531–540. https://doi.org/10.1080/09593339209385181

33. Prado R, García R, Rioboo C, Herrero C, Abalde J, Cid A (2009) Comparison of the sensitivity of different toxicity test endpoints in a microalga exposed to the herbicide paraquat. Environ Int 35:240–247. https://doi.org/10.1016/j.envint.2008.06.012

34. Reavie ED, Cangelosi AA, Allinger LE (2010) Assessing ballast water treatments: evaluation of viability methods for ambient freshwater microplankton assemblages. J Gt Lakes Res 36:540–547. https://doi.org/10.1016/j.jglr.2010.05.007

Chapter 14

Optimizing FlowCam Imaging Flow Cytometry Operation for Classification and Quantification of *Microcystis* Morphospecies

Dmitry Malashenkov, Veronika Dashkova, Ivan A. Vorobjev, and Natasha S. Barteneva

Abstract

Microcystis is a globally known cyanobacterium causing potentially toxic blooms worldwide. Different morphospecies with specific morphological and physiological characters usually co-occur during blooming, and their quantification employing light microscopy can be time-consuming and problematic. A benchtop imaging flow cytometer (IFC) FlowCam (Yokogawa Fluid Imaging Technologies, USA) was used to identify and quantitate different *Microcystis* morphospecies from environmental samples. We describe here the FlowCam methodology for sample processing and analysis of five European morphospecies of *Microcystis* common to the temperate zone. The FlowCam technique allows detection of different *Microcystis* morphospecies providing objective qualitative and quantitative data for statistical analysis.

Key words *Microcystis*, Phytoplankton, Imaging flow cytometry, FlowCam, Cyanobacteria, CyanoHABs

1 Introduction

The cyanobacterial genus *Microcystis* Kützing ex Lemmermann 1907 is one of the most important and well-known cyanobacteria due to the ability of some species to cause heavy water blooms in freshwater bodies [1–4], coastal marine [5–7], and estuarine regions [8–12]. It is noted for the mass development in lentic freshwaters facing eutrophication [13, 14] and is often responsible for harmful cyanobacterial blooms (CyanoHABs) in lakes, ponds, and drinking water reservoirs all over the world [4, 15, 16] except Antarctica [17].

This genus includes several species able to produce hepatotoxins and microcystins [16, 18–20], which are relatively stable in the water column and tend to accumulate in the ecological food chain,

Natasha S. Barteneva and Ivan A. Vorobjev (eds.), *Spectral and Imaging Cytometry: Methods and Protocols*,
Methods in Molecular Biology, vol. 2635, https://doi.org/10.1007/978-1-0716-3020-4_14,
© Springer Science+Business Media, LLC, part of Springer Nature 2023

causing animal poisoning and deaths [21, 22]. These cyanobacterial secondary metabolites are highly toxic [13, 14] and potentially dangerous to human health [23–28]. It is anticipated that the distribution and frequency of *Microcystis* blooms will be expanded under global climate change conditions [29–31].

Although under laboratory culture conditions most *Microcystis* strains abide as unicellular organisms [32–34], they tend to aggregate in typical colonies in the field [35, 36]. These micro- or macroscopic colonies are packed with irregularly arranged spherical cells in common, fine, colorless mucilage [37, 38]. Cells divide into three perpendicular planes and contain gas vesicles [37], one of the main criteria for identifying *Microcystis* [39], which control the buoyancy of colonies in the water column [38].

Such morphological characteristics (gas vesicles, cell size, colony shape and size, aggregation of cells in colonies, mucilage attributes, etc.) along with geographical distribution were used for the taxonomic identification of species according to the traditional botanical approach [40]. However, *Microcystis* natural colonies can be very variable in these attributes, especially at the beginning and end of the vegetation period [38], and the features of many populations do not correspond strictly to the species' limiting criteria [41]. Such phenotypic plasticity has raised doubts about the valid taxonomy of *Microcystis* [34, 41, 42].

Although the genus *Microcystis* is clearly delimited from other cyanobacterial genera by molecular sequencing [43], the taxonomy within the genus remains controversial since traditional botanical species of *Microcystis* were not recognized as well delimited clusters [41, 42, 44, 45]. Therefore, *Microcystis* populations with similar phenotypic characteristics (cell size, colony shape, size and density, mucilage features) previously classified as traditional species are now recognized as morphotypes or morphospecies [38, 46, 47]. The identification of these morphospecies with distinct phenotypic and ecophysiological properties is crucial for ecological and ecotoxicological studies [38].

Komárek & Anagnostidis have described 10 morphospecies from European water bodies and about 20 morphospecies from subtropical and tropical zones [37]. Šejnohová & Maršálek have divided the most commonly occurring European morphospecies (*M. ichtyoblabe, M. flos-aquae, M. aeruginosa, M. novacekii, M. viridis, M. wesenbergii*) into three cell-size clusters defined by a combination of morphological and molecular markers – small, close to *M. ichthyoblabe* (incl. *M. flos-aquae*), middle, based on *M. aeruginosa* (incl. *M. novacekii*); large, represented by *M. wesenbergii* [47].

Several *Microcystis* morphospecies such as *M. aeruginosa, M. wesenbergii, M. flos-aquae, M. ichtyoblabe, M. novacekii, M. smithii* may coexist and/or successively alternate during blooming [48–52]. Meanwhile, optimal environmental conditions and colony formation mechanisms for each morphospecies seem to be species-specific [53]. Water temperature and mixing appear to be the main factors controlling the succession of *Microcystis* morphospecies [49, 54]. Also, differences in toxin-producing, specifically in the compound of secreted oligopeptides, and toxicity strength between morphospecies were found in previous studies [37, 55–61].

In the presence of heavy blooms with co-dominance of several *Microcystis* morphospecies, it can be problematic to distinguish and count different colonies by means of routine light microscopy. In terms of health risks, it can endanger the public, as any delay in issuing health warnings will contribute to the risk of exposure [62]. In this case, the use of imaging flow analyzer FlowCam (Yokogawa Fluid Imaging Technologies, USA) appears to be feasible for this purpose. It counts and snapshots the particles up to 2 mM in size, which can be characterized by over 40 measurements, including length, width, aspect ratio, equivalent spherical diameter (ESD), area-based diameter (ABD), and others [63]. Being developed for oceanic plankton studies, FlowCam has attained success in plankton research and water management [64], but it came into use for the identification of colonial cyanobacteria in field freshwater samples [65] and estimation and enumeration of *Microcystis* cells and colonies [62, 66–68] only in recent times.

2 Materials

2.1 Instrumentation and Accessories

1. Benchtop FlowCam VS-4 imaging flow cytometer (Yokogawa Fluid Imaging Technologies, USA) equipped with 532 nM excitation laser (or alternative 488 excitation laser), color digital camera, and 4 magnification options ($2\times$, $4\times$, $10\times$, $20\times$) were used for *Microcystis* spp. analysis. Environmental samples containing *Microcystis* spp. were recorded in the laser-trigger mode to capture only fluorescent particles and using combinations of $10\times$ objective/100 μM flow cell and $20\times$ objective/50 μM flow cell for quantification and taxonomic identification (*see* **Note 1**).

2. Calibration beads (for auto-focusing).

3. Graduated Pasteur pipettes.

4. Funnels and cylinders.

2.2 Sample Preparation

Collected samples were fixed with 1% glutaraldehyde and stored in a fixative until the analysis. Before analysis on FlowCam imaging flow cytometer (IFC), samples were diluted with dH_2O in 1:4 to 1:1 ratio (depending on the sample cell concentration) and were filtered through 100 μM – sized mesh to avoid clogging of the 100 μM flow cell with dense colonies.

3 Methods

The whole process of FlowCam analysis consists of two parts: sample processing and analysis of acquired data consisted of library creation and classification using VisualSpreadsheet software. The classified data can then be exported in Excel format and used in other statistical and graphical software (Fig. 1).

3.1 Sample Processing

1. Open VisualSpreadsheet software, and select 10× magnification if using 100 μM flow cell (Fig. 2).

2. Adjust focus either using an auto-focus procedure or manual focusing (*see* **Note 2**). (For manual focusing, load 1 mL of a sample of interest and go to "Setup" tab to select "Autoimage (No save mode)". Manually adjust the position of the flow cell holder on the rail while running the sample in autoimage mode. Tighten the rail lock thumb screw when the particles are well focused.

3. Once the focus is adjusted, close the Optics door properly and go to "Setup" tab again to test trigger mode via "Trigger mode (No save)". Make sure that laser has been turned on via the laser safety key. If the laser is working, you can observe the laser light path inside the Optics compartment from above, and fluorescent particles will be imaged by a camera.

4. Go to "Context" tab to adjust the run settings.
   In the Capture tab:
   - Set distance to nearest neighbor (μM) 10 under the "Particle Capture".
   - Select both dark and light pixels under the "Particles defined by".
   - Select boxes to save collage and binary images under the "Images". The raw images can also be saved; however, they occupy large memory space.
   - Set the correct threshold (*see* **Note 3**).

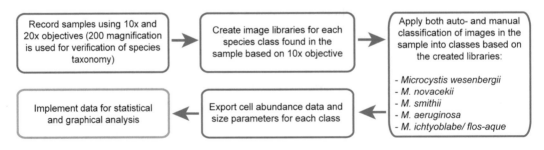

Fig. 1 FlowCam workflow scheme

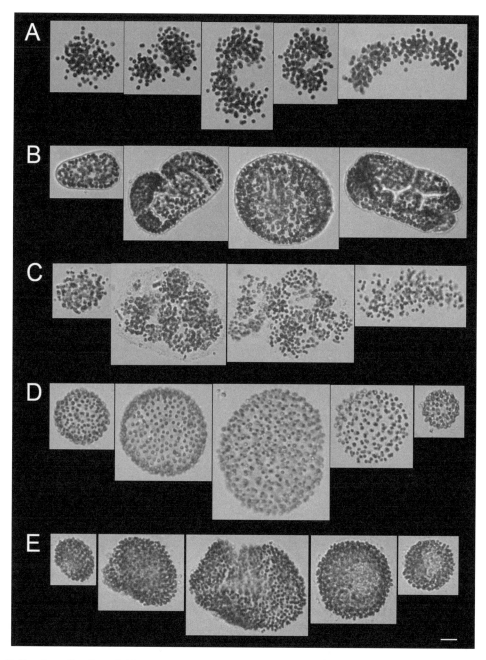

Fig. 2 FlowCam libraries of *Microcystis* morphospecies, 10× objective, 100 μM flow cell. (**a**) *Microcystis novacekii*; (**b**) *M. wesenbergii*; (**c**) *M. aeruginosa*; (**d**) *M. smithii*; (**e**) *M. flos-aquae/ichthyoblabe*. Scale bar = 20 μM

In the Fluidics tab:

– Select sample volume to be analyzed under the 'Settings".

– Tick the sample dilution box and enter the dilution ratio under the "Sample dilution".

In the Filter tab:

– Make sure that minimum and maximum particle diameter measurements are selected based on either ESD or ABD.

5. Go to Documents and create a folder where LST files obtained will be saved.

6. Load required sample volume using a volumetric cylinder or Eppendorf pipette into the sample funnel and go to "Trigger" tab to start a run. Transfer sample carefully to avoid formation of bubbles. If bubbles appear, use reverse pump flow function in the "Setup and Focus" window to push the bubble out of the sample.

7. Enter sample information in the popped-up window (e.g. sample name, date of collection, place of collection, fixed or live, etc.) and click OK. Enter the short sample name, select the designated folder from the list, and click OK.

8. To control the quality of images taken during the run (*see* **Note 4**).

9. After analyzing the sample volume set by the researcher, the run will stop automatically.

3.2 Data Classification

1. To view acquired data, go to "File" tab and "Open list" to select the LST file within the sample folder. Then click "Open View" tab to view the image collages.

2. Leaving the View window opened, go to the "File" tab in the main software window and click "Open Library Window". In the library window, go to "File" and click "New library", enter the library name and click OK. Go back to the opened View window and select individual images of *Microcystis* morphospecies to be added to the library. Right-click on one of the selected images and press "Add to library" and select the appeared name of the opened library. Go to "File" in the library window and click Save to save the changes made to the library. For statistical power, a minimum number of 40–50 images is recommended to be saved in one library.

3. It is also possible to add images of the same species to the created library from other samples. For this, open another sample as explained in 7 and add selected images to the opened library as explained in 8.

4. Create libraries for other *Microcystis* morphospecies using **steps 8–9**.

5. Once libraries are created, it is necessary to create a library-based statistical filter for further classification. Open one of the created libraries and go to "Filter" in the Library window and select "Build statistical filter". In the Filter dialog select "All relevant fields (From Ideal Particle)", click Save and enter the name of the filter (usually it is the same as the library name) in the appeared window and click Save. A combination of statistical parameters can be optimized to improve the classification.

6. Open the required sample file as in 7. Go to "File" tab in the main software window and click "Open Classification Window". This will open a separate classification window. Go to "File" tab in the classification window and select "Open Classification", where you can create a new classification. Enter the classification name and click Save. In the created classification, go to Classes tab and select New Class to start adding the classes. Enter the class name and click Add to select the corresponding statistical filter created earlier and click OK.

7. Create a required number of classes following instructions in 12 and select "Save Classification as template" in the "File" tab. This classification template can now be used for similar samples.

8. Once all classes are added, go to "Operation" tab and click "Run Auto Classification" to classify the sample. Most probably, the sample will not be properly classified due to the high degree of similarity among *Microcystis* morphospecies, and therefore additional manual sorting is required.

4 Notes

1. FlowCam analysis of large cell colonies such as *Microcystis* spp. requires a compromise between taxonomic resolution power provided by 20× and sample processing capacity provided by 10× and 4× objectives (Fig. 3). Images of *Microcystis* acquired using 4× objective provide low resolution making it difficult to identify different morphospecies. In comparison to 10×, 20× objective provides better image resolution enabling more accurate identification of the morphospecies. However, the depth of the flow cell associated with 20× objective limits the analysis of colonies larger than 50 µM and, therefore, cannot be used for quantitative analysis. We recommend using primarily 10× objective for quantitative characterization of Microcystis as it provides good image resolution for colonies up to 100 µM (average size), large sample processing capacity, fewer clogging events, and preserves the wholeness of the colonies that are also important for morphospecies identification. The use of 20× objective is still necessary to obtain a better representation of cell size and cell properties required for taxonomic identification.

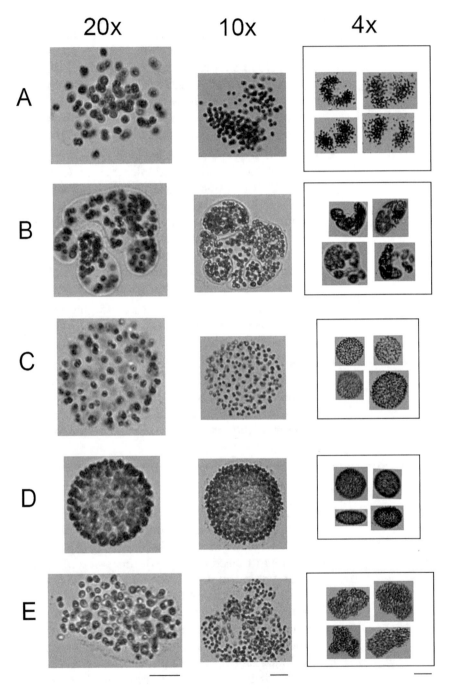

Fig. 3 FlowCam images of *Microcystis* morphospecies acquired using 4×, 10× and 20× objectives. (**a**) *Microcystis novacekii*; (**b**) *M. wesenbergii*; (**c**) *M. smithii*; (**d**) *M. flos-aquae/ichthyoblabe*; (**e**) *M. aeruginosa.* Scale bar = 20 μM

2. There are two options for focus adjustment available in Flow-CAM VS-4 series: auto-focusing and manual focusing. Auto-focus procedure is conducted using a bead solution of known size (20 µM beads for 10× and 20× objectives, 50 µM beads for 4× and 2× objectives). The auto-focus procedure is easier performed, provided the uniformity and clearness of beads enabling to achieve of less-subjective focus. However, the downside of auto-focusing is a washing procedure with substantial water volume required to wash out the beads from the sample funnel and flow cell afterward. If not washed properly, the beads will appear among recorded images and will affect the value of the total concentration of particles in the sample. The focus can also be adjusted manually without bead solution and based on the particles in the sample. Although it requires some practice of the operator, manual focusing enables to focus directly on the particles in the sample and skipping the after-bead washing step.

3. Different thresholds for dark and light pixels control the particle area in the binary image (Fig. 4). Binary images are used for acquiring particle dimensions such as area, area-based diameter, volume, etc., and therefore directly influence the size measurements produced by the software. Default threshold settings for dark and light pixels (Dark: 20; Light: 20) should be adjusted depending on the sample, particles of interest, and objective used to obtain accurate size measurements. We found that a combination of both dark and light pixels should be used for 10× objective-based analysis. Specifically, adjusting the threshold for dark pixels to 40–50 resulted in the more accurate contouring of dark *Microcystis* spp. colonies, and the appearance of spaces between cells in the colony on binary images compared to the default settings (Fig. 4).

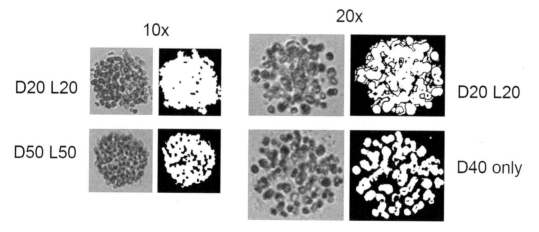

Fig. 4 Comparison of brightfield and binary images obtained for 10× and 20× objectives with different threshold settings

When using both dark and light pixels with default settings for a 20× objective, extra pixels are included in the binary image due to the light scattering resulting from the cell membrane of colonial cells (Fig. 4). Therefore, the use of light pixels in a 20× objective-based analysis might be inappropriate, and we recommend using only dark pixels with the threshold set to 30–40 for *Microcystis* sp. analysis.

4. It is important to constantly watch the quality of images taken during the run to avoid imaging of non-target particles that stick to the flow cell surface or large particles trapped in the flow cell.

   (a) If the focus has not been set up properly, other non-target particles that tend to accumulate on the flow cell surface may come into focus and get imaged by the camera. In this case, without stopping the run, pressing "Recalibrate" button in the Camera window may help to bring target particles into focus.

   (b) In some cases, large colonies may get trapped in the flow cell and reside in the field of view. If termination of the run and cleaning the flow cell is not possible, then try to pinch the tubing between the flow cell and the pump several times to displace the particle from the field of view.

5. When dealing with samples containing a high concentration of bloom-forming species such as *Microcystis* sp., clogging of the flow cell may be unavoidable.

   (a) To minimize clogging, wash the flow cell with an adequate volume of distilled H_2O or ethanol between samples. The pump flow can be set to 1–3 mL/min to increase cleaning efficiency.

   (b) It is also helpful to reverse the pump flow to clean the top of the flow cell by clicking "Go reverse" in the "Setup and Focus" window.

   (c) If clogging appeared nonetheless and cannot be cleaned by reverse flow, remove the flow cell with attached tubing and place it upside down by connecting the tubing ends to the sample container and pump. Set up the pump velocity to 1 mL/min in the Setup and Focus window and flush the flow cell with distilled H_2O or ethanol until the clog is removed. Instead of using continuous water flow, try to alternate a small quantity of water with air spacing to increase the force.

6. The data on the different classified morphospecies, including the concentration of colonies in the sample and size dimensions (length, diameter, area), can be obtained for a single image in Excel table format.

Acknowledgments

Work was supported by grant MES (Ministry of Education and Sciences, Kazakhstan) AP14872028 to N.S.B. and grant AP14860015 MES Kazakhstan to I.A.V. We are grateful to Harry Nelson (Yokogawa Fluid Technologies, USA) for his valuable advice with FlowCam instrumentation, and to Adina Zhumakhanova for help and discussion.

References

1. Reynolds CS (1987) Cyanobacterial water blooms. In: Callow JA (ed) Advances in botanical research, vol 13. Academic Press, London, pp 67–143

2. Dokulil MT, Teubner K (2000) Cyanobacterial dominance in lakes. Hydrobiologia 438:1–12. https://doi.org/10.1023/A:1004155810302

3. Oliver R, Ganf G (2000) Freshwater blooms. In: Whitton B, Potts M (eds) The ecology of cyanobacteria: their diversity in time and space. Kluwer Academic Publishers, pp 149–194

4. Harke MJ, Steffen MM, Gobler CJ et al (2016) A review of the global ecology, genomics, and biogeography of the toxic cyanobacterium, *Microcystis spp.* Harmful Algae 54:4–20. https://doi.org/10.1016/j.hal.2015.12.007

5. Mazur-Marzec H, Browarczyk-Matusiak G, Forycka K, Kobos J, Pliński M (2010) Morphological, genetic, chemical and ecophysiological characterization of two *Microcystis aeruginosa* isolates from the Vistula Lagoon, southern Baltic. Oceanologia 52:127–146

6. Belykh OI, Dmitrieva OA, Gladkikh AS et al (2013) Identification of toxigenic *Cyanobacteria* of the genus *Microcystis* in the Curonian Lagoon (Baltic Sea). Oceanology 53:71–79. https://doi.org/10.1134/S0001437013010025

7. Prasath B, Kumar N, Thillainayagam SP (2014) First report on the intense cyanobacteria *Microcystis aeruginosa* Kützing, 1846 bloom at Muttukkadu Backwater, Southeast coast of India. Indian J Mar Sci 43:258–262

8. Lehman PW, Boyer G, Hall C, Waller S, Gehrts K (2005) Primary research paper distribution and toxicity of a new colonial *Microcystis aeruginosa* bloom in the San Francisco Bay Estuary, California. Hydrobiologia 541:87–99. https://doi.org/10.1007/s10750-004-4670-0

9. Moisander PH, Lehman PW, Ochiai M, Corum S (2009) Diversity of *Microcystis aeruginosa* in the Klamath River and San Francisco Bay delta, California, USA. Aquat Microb Ecol 57:19–31

10. Vasudevan S, Arulmoorthy MP, Gnanamoorthy P, Ashokprabu V, Srinivasan M (2015) Continuous blooming of harmful microalgae *Microcystis aeruginosa* Kutzing, 1846 in Muttukadu estuary, Tamilnadu, southeast coast of India. IJSIT 4:15–23

11. Preece EP, Hardy FJ, Moore BC, Bryan M (2017) A review of microcystin detections in estuarine and marine waters: environmental implications and human health risk. Harmful Algae 61:31–45. https://doi.org/10.1016/j.hal.2016.11.006

12. Kurobe T, Lehman PW, Hammock BG, Bolotaolo MB, Lesmeister S, Teh SJ (2018) Biodiversity of cyanobacteria and other aquatic microorganisms across a freshwater to brackish water gradient determined by shotgun metagenomic sequencing analysis in the San Francisco Estuary, USA. PLoS One 13:e0203953. https://doi.org/10.1371/journal.pone.0203953

13. Pflugmacher S, Wiegand C (2001) Uptake, enzyme effects and metabolism of microcystins-LR in aquatic organisms. In: Chorus I (ed) Cyanotoxins. Springer, Berlin, pp 249–260

14. Ouahid Y, Pérez-Silva G, del Campo FF (2005) Identification of potentially toxic environmental Microcystis by individual and multiple PCR amplification of specific microcystin synthetase gene regions. Environ Toxicol 20:235–242. https://doi.org/10.1002/tox.20103

15. Baker LC (2002) Cyanobacterial harmful algal blooms (CyanoHABs): developing a public health response. Lake Reserv Manage 18:20–31

16. Codd GA, Lindsay J, Young FM, Morrison LF, Metcalf JS (2005) Cyanobacterial toxins. In: Huisman J, Matthijs HCP, Visser PM (eds) Harmful cyanobacteria. Springer, Dordrecht, pp 1–23

17. Zurawell RW, Chen H, Burke JM et al (2005) Hepatotoxic cyanobacteria: a review of the biological importance of microcystins in freshwater environments. J Toxicol Environ Health B Crit Rev 8:1–37. https://doi.org/10.1080/10937400590889412

18. Park HD, Iwami C, Watanabe MF et al (1998) Temporal variabilities of the concentration of inter- and extracellular microcystin and toxic *Microcystis* species in a hypertrophic lake, Lake Suwa, Japan (1991–1994). Environ Toxicol Water Qual 13:61–72

19. Chorus I, Bartram J (eds) (1999) Toxic cyanobacteria in water. A guide to their public health consequences, monitoring and management. E&FN Spon, London

20. Sivonen K, Jones G (1999) Cyanobacterial toxins. In: Chorus I, Bartram J (eds) Toxic cyanobacteria in water. E & FN Spon, London, pp 41–111

21. Ferrão-Filho AS, Kozlowsky-Suzuki B (2011) Cyanotoxins: bioaccumulation and effects on aquatic animals. Mar Drugs 9:2729–2772

22. Pawlik-Skowrońska B, Kalinowska R, Skowroński T (2013) Cyanotoxin diversity and food web bioaccumulation in a reservoir with decreasing phosphorus concentrations and perennial cyanobacterial blooms. Harmful Algae 28:118–125

23. Carmichael WW, Azevedo SM, An JS et al (2001) Human fatalities from cyanobacteria: chemical and biological evidence for cyanotoxins. Environ Health Perspect 109:663–668

24. Fitzgerald DJ (2001) Cyanotoxins and human health - overview. In: Chorus I (ed) Cyanotoxins – occurrence, causes, consequences. Springer, Berlin, pp 179–190

25. Azevedo SMFO, Carmichael WW, Jochimsen EM et al (2002) Human intoxication by microcystins during renal dialysis treatment in Caruaru-Brazil. Toxicology 181-182:441–446

26. Hu Z, Chen H, Li Y, Gao L, Sun C (2002) The expression of bcl-2 and bax genes during microcystin induced liver tumorigenesis. Chin J Prev Med 36:239–242

27. Falconer IR (2007) Cyanobacterial toxins present in *Microcystis aeruginosa* extracts – more than microcystins! Toxicon 50:585–588. https://doi.org/10.1016/j.toxicon.2007.03.023

28. Žegura B, Volčič M, Lah TT, Filipič M (2008) Different sensitivities of human colon adenocarcinoma (CaCo-2), astrocytoma (IPDDC-A2) and lymphoblastoid (NCNC) cell lines to microcystin-LR induced reactive oxygen species and DNA damage. Toxicon 52:518–525

29. Paerl HW, Huisman J (2008) Blooms like it hot. Science 320:57–58

30. Michalak AM, Anderson EJ, Beletsky D, Boland S, Bosch NS (2013) Record-setting algal bloom in Lake Erie caused by agricultural and meteorological trends consistent with expected future conditions. Proc Natl Acad Sci USA 110:6448–6452. https://doi.org/10.1073/pnas.1216006110

31. Paerl HW, Otten TG (2013) Harmful cyanobacterial blooms: causes, consequences and controls. Microb Ecol 65:995–1010

32. Li L, Zhu W, Wang T, Luo Y, Chen F, Tan X (2013) Effect of fluid motion on colony formation in *Microcystis aeruginosa*. Water Sci Eng 6:106–116

33. Yang Z, Kong F, Shi X, Zhang M et al (2008) Changes in the morphology and polysaccharide content of *Microcystis aeruginosa* (*Cyanobacteria*) during Flagellate grazing. J Phycol 44:716–720

34. Xiao M, Li M, Reynolds CS (2018) Colony formation in the cyanobacterium *Microcystis*. Biol Rev Camb Philos Soc 93:1399–1420. https://doi.org/10.1111/brv.12401

35. Zhu W, Li M, Luo Y et al (2014) Vertical distribution of *Microcystis* colony size in Lake Taihu: its role in algal blooms. J Great Lakes Res 40:949–955

36. Rowe MD, Anderson EJ, Wynne TT et al (2016) Vertical distribution of buoyant *Microcystis* blooms in a Lagrangian particle tracking model for short-term forecasts in Lake Erie. J Geophys Res Oceans 121:5296–5314. https://doi.org/10.1002/2016JC011720

37. Komárek J, Anagnostidis K (1998) Cyanoprokaryota 1. Teil: chroococcales. In: Ettl H, Gartner G, Heynig H, Mollenhauer D, Fischer G (eds) Susswasserflora von Mitteleuropa 19/1. Springer, Stuttgart, pp 1–548

38. Komárek J, Komárková J (2002) Review of the European *Microcystis*-morphospecies (Cyanoprokaryotes) from nature. Czech Phycol Olomouc 2:1–24

39. Holt JG, Kreig NR, Sneath PHA, Staley JT, Williams ST (1994) Group 11. Oxygenic phototrophic bacteria. In: Hensyl WR (ed) Bergey's manual of determinative bacteriology, 9th edn. Williams & Wilkins, Baltimore, pp 377–425

40. Komárková J, Mugnai MA, Sili C, Komárek O, Turicchia S (2005) Stable morphospecies within the 16S rRNA monophyletic genus *Microcystis* (Kutzing) Kutzing. Algol Stud/Arch Hydrobiol 117:279–295

41. Otsuka S, Suda S, Li R, Matsumoto S, Watanabe MM (2000) Morphological variability of colonies of *Microcystis* morphospecies in culture. J Gen Appl Microbiol 46:57–68

42. Otsuka S, Suda S, Shibata S, Oyaizu H, Matsumoto S, Watanabe MM (2001) A proposal for the unification of five species of the cyanobacterial genus *Microcystis* (Kutzing) Lemmermann 1907 under the rules of the bacteriological code. Int J Syst Evol Microbiol 51:873–879

43. Li R, Yokota A, Sugiyama J, Watanabe M, Hiroki M, Watanabe M (1998) Chemotaxonomy of planktonic cyanobacteria based on non-polar and 3-hydroxy fatty acid composition. Phycol Res 46:21–28

44. Castenholz RW (2001) Oxygenic photosynthetic bacteria. In: Boone DR, Castenholz RW (eds) Bergey's manual of systematic bacteriology, vol 1, 2nd edn. Springer, New York, pp 473–600

45. Mituleţu M, Druga B, Hegedus A, Coman C, Dragos N (2013) Phylogenetic analysis of *Microcystis* strains (cyanobacteria) based on the 16S-23S ITS and cpcBA-IGS markers. Ann Roman Soc Cell Biol 18:22–31

46. Sanchis D, Padilla C, Del Campo FF, Quesada A, Sanz-Alferez S (2005) Phylogenetic and morphological analyses of *Microcystis* strains (*Cyanophyta/Cyanobacteria*) from a Spanish water reservoir. Nova Hedwigia 81: 431–448

47. Šejnohová L, Maršálek B (2012) Microcystis. In: Whitton BA (ed) Ecology of cyanobacteria II: their diversity in space and time, 2nd edn. Springer, Dordrecht, pp 195–227

48. Ozawa K, Fujioka H, Muranaka M et al (2005) Spatial distribution and temporal variation of *Microcystis* species composition and microcystin concentration in Lake Biwa. Environ Toxicol 20:270–276

49. Imai H, Chang K-H, Kusaba M, Nakano S (2008) Temperature-dependent dominance of *Microcystis* (*Cyanophyceae*) species: *M. aeruginosa* and *M. wesenbergii*. J Plankton Res 31:171–178. https://doi.org/10.1093/plankt/fbn110

50. Yamamoto Y, Nakahara H (2009) Seasonal variations in the morphology of bloom-forming cyanobacteria in a eutrophic pond. Limnology 10:185–193. https://doi.org/10.1007/s10201-009-0270-z

51. Hu L, Shan K, Lin L, Shen W, Huang L, Gan N, Song L (2016) Multi-year assessment of toxic genotypes and Microcystin concentration in northern Lake Taihu, China. Toxins 8:23. https://doi.org/10.3390/toxins8010023

52. Krstić S, Alesovski B, Komárek J (2017) Rare occurrence of nine *Microcystis* species (*Chroococcales, Cyanobacteria*) in a single Lake (Lake Dojran, fYR Macedonia). Adv Oceanogr Limnol 8. https://doi.org/10.4081/aiol.2017.6236

53. Xiao M, Willis A, Burford MA, Li M (2017) Review: a meta-analysis comparing cell-division and cell-adhesion in *Microcystis* colony formation. Harmful Algae 67:85–91

54. Li M, Xiao M, Zhang P, Hamilton DP (2018) Morphospecies-dependent disaggregation of colonies of the cyanobacterium *Microcystis* under high turbulent mixing. Water Res 141: 340–348. https://doi.org/10.1016/j.watres.2018.05.017

55. Fastner J, Erhard M, von Döhren H (2001) Determination of oligopeptide diversity within a natural population of *Microcystis* spp. (*Cyanobacteria*) by typing single colonies by matrix-assisted laser desorption ionization-time of flight mass spectrometry. Appl Environ Microbiol 67:5069–5076

56. Via-Ordorika L, Fastner J, Kurmayer R et al (2004) Distribution of microcystin-producing and non-microcystin-producing *Microcystis* sp. in European freshwater bodies: detection of microcystins and microcystin genes in individual colonies. Syst Appl Microbiol 27:592–602

57. Welker M, Marsálek B, Šejnohová L, von Döhren H (2006) Detection and identification of oligopeptides in *Microcystis* (cyanobacteria) colonies: toward an understanding of metabolic diversity. Peptides 27:2090–2103

58. Welker M, Šejnohová L, Harustiakova D, Döhren H, Jarkovský J, Marsálek B (2007) Seasonal shifts in chemotype composition of "Microcystis" sp. communities in the pelagial and the sediment of a shallow reservoir. Limnol Oceanogr 52:609–619. https://doi.org/10.4319/lo.2007.52.2.0609

59. Xu Y, Wu Z, Yu B, Peng X, Yu G, Wei Z, Wang G, Li R (2011) Non-microcystin producing *Microcystis wesenbergii* (Komárek) Komárek (cyanobacteria) representing a main waterbloom-forming species in Chinese waters. Environ Pollut 156:162–167. https://doi.org/10.1016/j.envpol.2007.12.027

60. Gan NQ, Huang Q, Zheng LL et al (2010) Quantitative assessment of toxic and nontoxic *Microcystis* colonies in natural environments using fluorescence in situ hybridization and flow cytometry. Sci China Life Sci 53:973–

980. https://doi.org/10.1007/s11427-010-4038-9

61. Liu Y, Tan W, Wu X, Wu Z, Yu G, Li R (2011) First report of microcystin production in *Microcystis smithii* Komárek and Anagnostidis (Cyanobacteria) from a water bloom in Eastern China. J Environ Sci 23:102–107

62. Graham M, Cook J, Graydon J, Kinniburgh D, Nelson H, Pilieci S, Vinebrooke RD (2018) High-resolution imaging particle analysis of freshwater cyanobacterial blooms: FlowCam analysis of cyanobacteria. Limnol Oceanogr Methods 16. https://doi.org/10.1002/lom3.10274

63. Poulton NJ (2016) FlowCam: quantification and classification of phytoplankton by imaging flow cytometry. In: Barteneva NS, Vorobjev IA (eds) Imaging flow cytometry: methods and protocols, Methods in molecular biology, vol 1389. Humana Press, New York, pp 237–247

64. Dashkova V, Malashenkov D, Poulton N, Vorobjev IA, Barteneva NS (2017) Imaging flow cytometry for phytoplankton analysis. Methods 112:188–200. https://doi.org/10.1016/j.ymeth.2016.05.007

65. Magonono M, Oberholster PJ, Addmore S, Stanley M, Gumbo JR (2018) The presence of toxic and non-toxic cyanobacteria in the sediments of the Limpopo river basin: implications for human health. Toxins 10:e269. https://doi.org/10.3390/toxins10070269

66. Wang C, Wu X, Tian C, Li Q, Tian Y, Feng B, Xiao B (2014) A quantitative protocol for rapid analysis of cell density and size distribution of pelagic and benthic *Microcystis* colonies by FlowCAM. J Appl Phycol 27:711–720. https://doi.org/10.1007/s10811-014-0352-0

67. Lehman PW, Kurobe T, Lesmeister S, Baxa D, Tung A, Teh SJ (2017) Impacts of the 2014 severe drought on the *Microcystis* bloom in San Francisco estuary. Harmful Algae 63:94–108. https://doi.org/10.1016/j.hal.2017.01.011

68. Park J, Kim Y, Kim M, Lee WH (2019) A novel method for cell counting of *Microcystis* colonies in water resources using a digital imaging flow cytometer and microscope. Environ Eng Res 24:397–403. https://doi.org/10.4491/eer.2018.266

INDEX

Natasha S. Barteneva and Ivan A. Vorobjev (eds.), *Spectral and Imaging Cytometry: Methods and Protocols*, Methods in Molecular Biology, vol. 2635, https://doi.org/10.1007/978-1-0716-3020-4, © Springer Science+Business Media, LLC, part of Springer Nature 2023

Printed in the United States
by Baker & Taylor Publisher Services